THE LEGO® MINDSTORMS® NXT ZOO!

THE LEGO® MINDSTORMS® NXT ZOO!

an unofficial, kid-friendly guide to building robotic animals with LEGO MINDSTORMS NXT

fay **rhodes**

Printed in the United States of America

11 10 09 08   2 3 4 5 6 7 8 9

ISBN-10: 1-59327-170-0
ISBN-13: 978-1-59327-170-1

Publisher: William Pollock
Production Editor: Megan Dunchak
Cover and Interior Design: Octopod Studios
Developmental Editor: William Pollock
Copyeditor: Nancy Sixsmith
Compositor: Riley Hoffman

For information on book distributors or translations, please contact No Starch Press, Inc. directly:

No Starch Press, Inc.
555 De Haro Street, Suite 250, San Francisco, CA 94107
phone: 415.863.9900; fax: 415.863.9950; info@nostarch.com; www.nostarch.com

Library of Congress Cataloging-in-Publication Data

Rhodes, Fay.
  The LEGO MINDSTORMS NXT zoo! : an unofficial, kid-friendly guide to building robotic animals with LEGO MINDSTORMS NXT / Fay Rhodes.
      p. cm.
  Includes index.
  ISBN-13: 978-1-59327-170-1
  ISBN-10: 1-59327-170-0
 1. Robotics--Juvenile literature. 2. LEGO toys--Juvenile literature. I. Title.
TJ211.2.R48 2008
629.8'92--dc22
                                      2008000316

To my husband, Rick, who introduced me to the LEGO MINDSTORMS NXT

brief contents

contents in detail

preface

Half of the LEGO® MINDSTORMS® NXT community is made up of brilliant adults who have a passion for building mind-boggling robots. These engineers, physicists, and other "rocket scientists" deserve credit for keeping the LEGO MINDSTORMS development and marketing teams on their toes.

But this book is written for the other half of the NXT community:

* Boys and girls who love animals
* Parents who want to prepare their children for a high-tech world
* Teachers who want to use emerging technology to engage a broad spectrum of students
* Artists who want to create art that moves and responds to its environment
* Anyone who is new to (or intimidated by) the LEGO MINDSTORMS NXT

There's been a concerted effort to make the building instructions in this book as simple as possible; but keep in mind that robotics isn't simple. Building any robot requires patience, perseverance and a bit of creativity.

Chapter 1 begins by telling you what you need to build these robots and where to find anything you are missing. If you're technologically challenged, you'll appreciate Chapter 2, which is a guide to basic terms and processes. If you're an experienced NXT user, just skip it.

The Hip-Hoppers in Chapter 3—Ribbit the frog and Bunny the rabbit—stand out for the fact that they actually hop. I have not seen hopping robots anywhere else. If you're wondering why I've included two robots that are so similar—well, my husband prefers the frog, and I prefer the rabbit, so I've included both.

Other animals include four-legged walkers—Sandy the camel in Chapter 4, Snout the alligator in Chapter 6, LEGOsaurus the dinosaur in Chapter 7, and Pygmy the pygmy elephant in Chapter 8—and Spiderbot, an eight-legged spider in Chapter 5.

Finally, there are my favorite two robots—Polecat the skunk and Strutter the peacock on wheels in Chapters 9 and 10—each of which responds to its environment in a unique way.

In case you need them, we've also included three appendices—enrichment ideas for teachers, troubleshooting tips, and some resources on the Web you might find helpful.

Of course, you'll learn as you build these robots, but most of all I hope you'll have fun building robots that look cool and work as promised.

This book would not have happened without the encouragement and good advice of Jim Kelly, NXT author and moderator of The NXT STEP blog. He is a generous person and a great asset to the NXT community.

If you can understand the instructions for building these robots, it is thanks to Rick and Connor Rhodes, who spent many long hours building and rebuilding each of the models and evaluating the instructions. If these animals are just the ones you'd like to build, you can thank Jake Olen, my unofficial child consultant, who unknowingly made my day when he suggested the animals I'd already built.

Thanks to Bill Pollock for the chance to publish these designs, to Megan Dunchak, the world's most patient and diplomatic production editor, and to Riley Hoffman for making it all look good.

Finally, I would like to thank God for my talents, gifts, and opportunities. His plans are always so much better than my own.

the right stuff:
do you have what it takes to build "living" robots?

Before you plunge in and build the robots in this book, you'll need to determine whether you have the parts to build them. If you have the NXT Base Set (Education or retail version) plus the Education Resource Set, you are all set to build three of the models in this book: the Hip-Hoppers (Bunny and Ribbit) and Polecat. The rest can be built with some part substitutions or a few additional parts—parts that you will use again and again as you design your own robots.

Each model includes a *Bill of Materials (BOM)* that lists exactly which parts you will need to build it, because nothing is more disappointing than getting midway through the instructions, only to learn you are missing some parts.

This book's companion website (*http://www.thenxtzoo.com/*) offers some at-a-glance charts for your information. These charts show which parts are used in each robot, which parts are included in each base set, and what is included in the Education Resource Set. You'll also find a chart listing the parts required to build each of the models.

the starting point

My first assumption is that you have either the LEGO MINDSTORMS NXT Education Base Set (Education Base Set) or the retail LEGO MINDSTORMS NXT Invention Kit (retail kit). If you have only an Education Base Set, you will need to expand your NXT parts stash in order to build these robots. Too, the Education Base Set and the retail kit do not contain the same parts. Both sets contain one NXT brick, three NXT motors, and seven power cables, which are all the electrical parts that you need, but neither contains all the beams, axles, and connectors that you'll need.

Fortunately, all the parts used in these models are readily available, although you might need to buy them as part of a set. A good source of individual parts is BrickLink (*http://www.bricklink.com/*), which is an eBay-like place for LEGO parts.

some parts are recommended, not required

In my building instructions, I note that some parts are not required, but were added to enhance the appearance of your robot (for example, to make Ribbit look more like a frog, Pygmy more like an elephant, and so on).

Other enhancements include using colored pieces, which you can get by buying TECHNIC kits or through BrinkLink. For example, the chest and neck of Strutter the peacock use blue parts. I happened to have a TECHNIC kit with blue parts, and using them really enhanced the look of the model. Of course, blue parts aren't necessary, but they really add to the effect.

Another example is the use of black pieces for Polecat. The LEGO online store offers all-black supplementary beam packs (see Figure 1-4), so this should be an easy enhancement.

the education resource set—recommended for everyone

Some of the robots in this book have parts that are only found in the Education Resource Set, which is sold by LEGO Education (*http://www.legoeducation.com/*). For example:

* Wheels for Sandy the Camel's feet
* Dart and cannon for Polecat's projectile
* Large wheels used on Strutter and Polecat
* Tusks on Pygmy

In addition to offering a wide variety of additional basic parts that you will need to build your robots (pins, beams, gears, and so on), the Education Resource Set also offers enhancements like yellow bushings, which I use to make distinctive (and more noticeable) eyes. (You can find a complete list of what is included in the Education Resource Set at *http://www.peeron.com/.*)

Although you might be able to purchase the parts from the Education Resource Set individually through BrickLink, I strongly recommend that you buy the kit itself. No matter which base set you own, the Education Resource Set substantially expands your NXT building options—and it comes in its own plastic storage container.

hassenpins—an NXT favorite

Hassenpins (shown in Figure 1-1) are used in most of the models but are not included in the Education Base Set; they're in the Education Resource Set.

3 × 5 L shape with quarter ellipse beams

This *L*-shaped piece (Figure 1-2) is merely cosmetic on Ribbit, but it is required for Spiderbot. It is found in the Education Base Set, and there are only four in each set. If you have two Education Base Sets, you're in business. Owners of the retail kit will need to acquire these beams elsewhere. There are a number of possible sources, as the next section explains.

Figure 1-1: The Hassenpin is officially called TECHNIC Beam 3 × 3 Bent with Pins.

Figure 1-2: The 3 × 5 L Shape with Quarter Ellipse beam

where to find parts

Here's a list of resources you may consider using in your quest for parts.

peeron

Peeron (*http://www.peeron.com/*) is a tremendous resource of information about anything LEGO. For example, Figure 1-3 shows a page from Peeron.com with information about the piece I use to support the legs in Spiderbot: the TECHNIC Liftarm 3 × 5 L Shape with Quarter Oval. (Note that Peeron sometimes uses a slightly different name for a part. In this example, LEGO uses *ellipse*, while Peeron uses *oval*.)

This page tells you which LEGO kits include this part, which online stores have it in stock, and how much they are charging for it. Clicking a price link will take you directly to a page through which you can buy the part.

To use this Peeron feature, search for the part name or a portion of the name. If you don't know the name of the part, type *NXT* in the search box. This will show you all the NXT sets. If you click the number next to the title of a set, you will be shown a list of parts contained in the set (with pictures). When you find the right name, Peeron will take you directly to the page for that part, like the page shown in Figure 1-3.

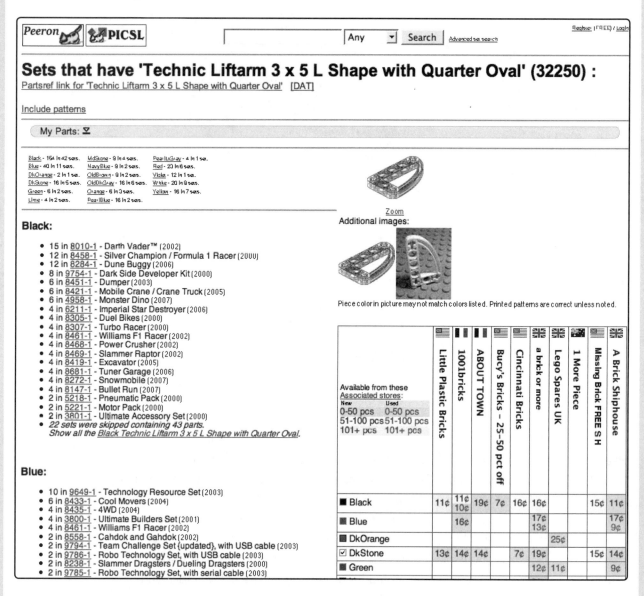

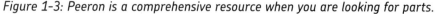

Figure 1-3: Peeron is a comprehensive resource when you are looking for parts.

pick-a-brick

Some of the parts you might need are available for pennies apiece from the Pick-A-Brick section of the LEGO online store (found at *http://www.lego.com/*), including the following:

* Bion Eye (a white version of LEGOsaurus's armor plates)
* Black pins
* Gray pins
* Blue axle pins
* Tan axle pins

If you're like me, some of these pins have gone into hiding around your work area, so you can't go wrong if you order a good supply of extra pins.

TECHNIC accessories

Currently, supplemental beam packs (listed under TECHNIC parts and accessories at the LEGO online store) are about $10 each and are well worth the cost. Figure 1-4 shows what this pack includes.

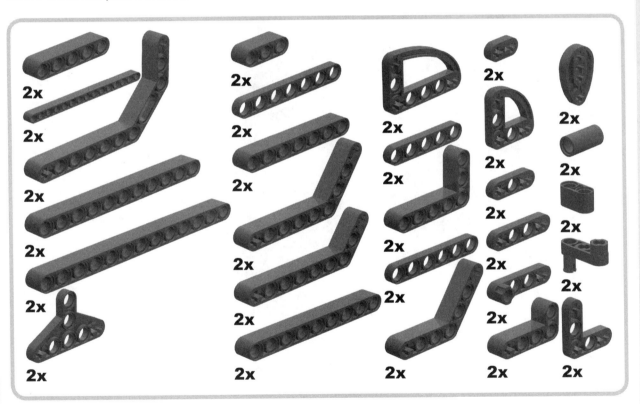

Figure 1-4: Pack of TECHNIC beams available at the online LEGO store

Building "living" robots always requires a variety of gears, and usually more than are provided in the base sets. Figure 1-5 shows a pack of gears currently available from the LEGO Online Store for about $13.

Figure 1-5: Pack of gear wheels available from the LEGO online store

mayan adventure pack

This package of parts was assembled by LEGO Education to accompany Jim Kelly's book, *The Mayan Adventure*, but anyone can purchase it. As of this writing, it is available for about $7, even without purchasing the book. To my knowledge, this is currently the only way you can buy the Hassenpins outside of a retail set or an Education Resource Set. Figure 1-6 shows the parts included in the Mayan Adventure Parts Pack.

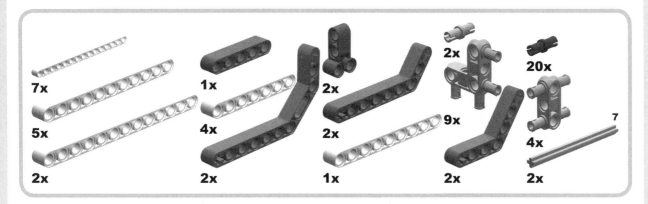

Figure 1-6: This parts pack was created to provide supplemental parts to readers of The Mayan Adventure.

companion website

Finally, this book's companion website, *http://www.thenxtzoo.com/*, offers up-to-date information on parts for these models (including an at-a-glance chart) as well as corrections to the book. If you're missing some parts, you'll find that some can be replaced with alternate pieces. The companion website offers some suggestions and guidelines for substituting parts. The substitutions should work as well as the parts listed in the building instructions.

the zookeeper's guide:

what you need to know to create the animals in the NXT zoo

Sometimes experienced NXT users take for granted the meanings of various terms, methods, and procedures. I offer this information for the new user, to make building successful robots as easy as possible.

coast vs. brake

There are times when you will be offered the option of *Brake* or *Coast* (shown at the lower right of Figure 2-1, beside the words *Next Action*). If you choose *Brake*, the robot will stop moving—like when you step on the brakes hard. If you choose *Coast*, the motors will stop turning, but the robot will stay in motion until it eventually stops—like when you coast on a bike. If you always choose *Brake*, your battery will drain faster. When it is not actually necessary for a motor to stop on a dime, choose *Coast*.

Figure 2-1: A sample configuration panel

direction

One of the most confusing things about programming a robot is making sure the program you've written moves the motors in the direction you want them to move. The motors can be built into your

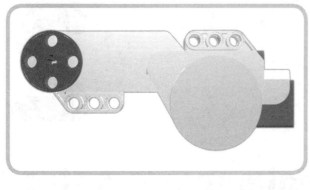

Figure 2-2: In the actual motor, the rotating part is orange.

robots in so many different ways! In NXT-G programming, the up arrow (↑) is referred to as *forward* and the down arrow (↓) is referred to as *backward*, but in many robots you'll need to choose forward to make your robot go backward. For example, hold a motor as shown in Figure 2-2. The orange part is the part that turns. In this position, clockwise motion results when you choose forward (↑), and counterclockwise motion is the result when you choose backward (↓).

downloading a program from your computer to the NXT brick

This is the procedure for getting an NXT-G program from your computer to your NXT brick:

1 Connect the computer to the NXT brick with the USB cord.
2 Turn on the brick.
3 Click the **Download** button (shown in Figure 2-3) and wait for the beep.

Figure 2-3: The lower left button downloads your program.

changing inches to centimeters

When you want to change from the default distance setting of inches to centimeters, the computer will make the calculation for you, but it is easy to get wrong if you don't enter the changes carefully. Here is the procedure for changing the distance to a specific number of centimeters:

1 Click the pull-down menu next to *Inches* (seen at the lower right of Figure 2-4) and select **Centimeters**. It is important that you do this first.
2 Type the number of centimeters you want.

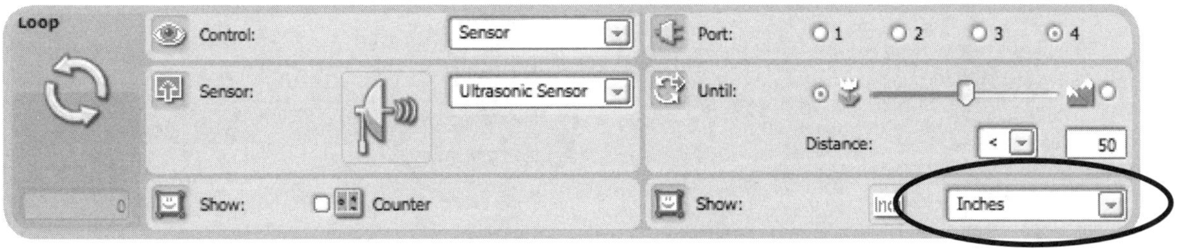

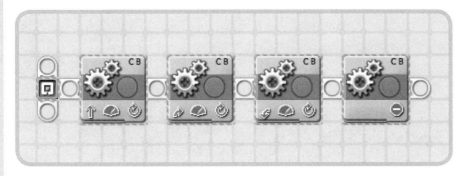

Figure 2-4: A configuration panel showing Inches at the lower right

my blocks

A long, complicated program is easier to understand and write when you break it into smaller parts. In the NXT-G program, we do this by creating miniprograms called *My Blocks*. Once they are created, My Blocks can be used to build other programs as well.

To create a My Block, first create the small program you want to convert to a My Block. In Figure 2-5, I have created a program with four different Move blocks.

Figure 2-5: These four blocks will become a small program that moves the C and B motors forward, turns them side to side, and then stops them.

Next, use your mouse to select all the blocks in the program as seen in both Figures 2-5 and 2-6. (The borders of selected blocks are turquoise when selected.)

Figure 2-6 shows where the *Create My Block* button is at the top of the screen. Click this button.

You will then see a screen like Figure 2-7. Use this screen to name your My Block. (I've named this My Block *wiggle*.) Once you've typed a name, click the **Next** button at the bottom.

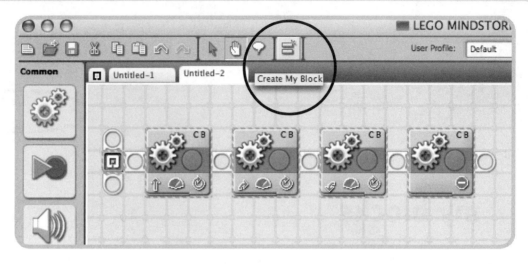

Figure 2-6: The Create My Block button is found on the top toolbar.

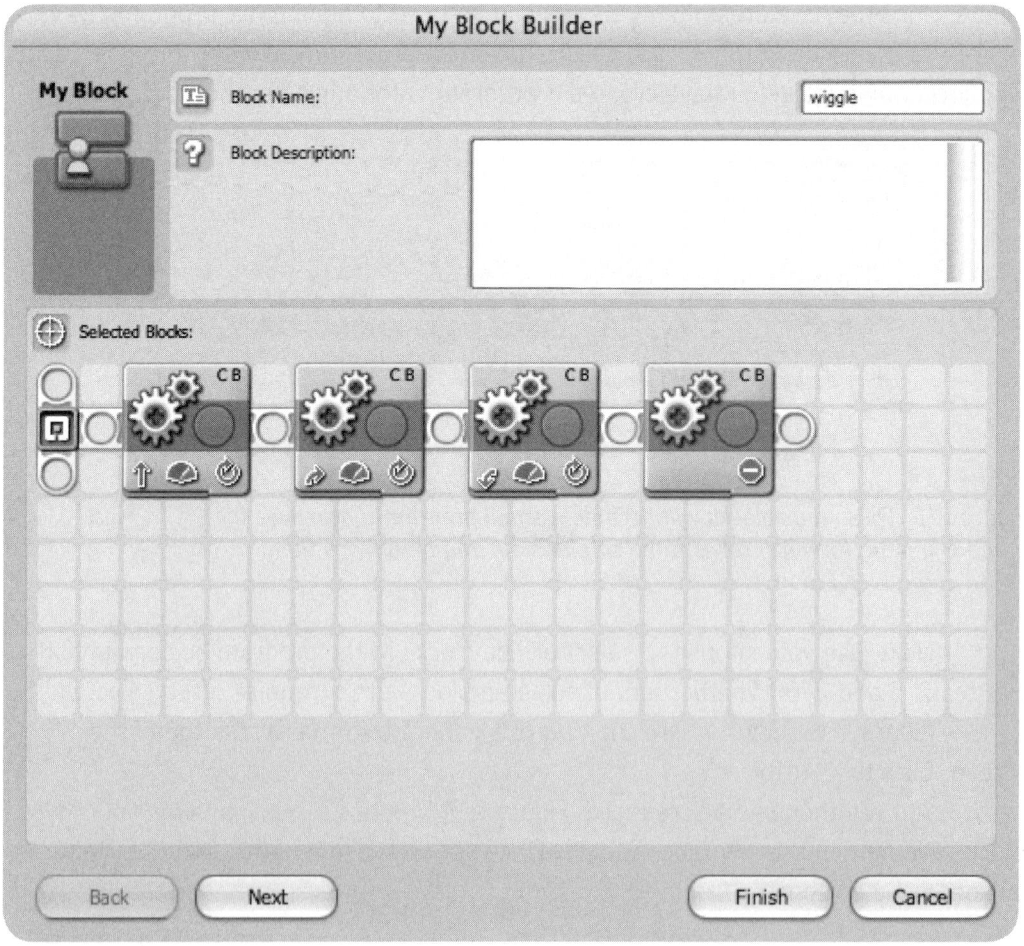

Figure 2-7: Name the block using this screen.

After you click Next, a screen will appear like the one shown in Figure 2-8. On this page, you can drag icons onto the open square at the top. This is not required, but it will help you to recognize your different My Blocks at a glance.

Figure 2-8: This screen allows you to give your icon a unique look.

Click **Finish** to complete the My Block.

pins

Choosing the correct pins will make a significant difference in the success of your robots. There will be times when it does not matter which color you use, but remember this:

* Yellow and gray pins are for use in joints that will move, such as the ones connecting legs to gears.
* Black and blue pins hold more firmly and are less likely to allow movement. Use them when you want something to hold tightly, like the spikes on LEGOsaurus.

port selection

By the same token, be sure to use the correct ports. Motors can use ports A, B, and C. Sensors can be plugged into ports 1, 2, 3, and 4.

Also, make sure that the ports you select are the same ones referenced in your programs.

power settings

While I will give you power settings to use for all the robots in this book, the correct settings for your robot will depend on the charge in your batteries. If your batteries are getting low, your robot might have difficulty moving.

Power settings will also be affected by the surface your robot will move on and any extra weight you might add to it, such as feathers on Strutter or fur on Polecat's tail. Each of these factors can significantly affect your robot's ability to move.

If your robot is quickly falling apart or falling over, try reducing the power first.

 NOTE This *flip!* symbol means that the model has been turned around since the last step.

3

hip-hoppers: an NXT frog and rabbit

When I design any animal robot, the first thing I do is observe the animals I want to imitate—in this case, frogs and rabbits. I look at photos, trace the forms of their skeletons, and observe how those skeletons move.

The next step is to open my NXT stash and contemplate how I might build the robot. In this case, I couldn't find instructions for any other hopping robots, so I had to figure it out myself. The results are simple but effective.

After building these two hoppers, maybe you'll be ready to tackle something bigger—like a kangaroo, for example. If you do, consider posting it on NXTLOG on the LEGO website.

NOTE **Running a hopping robot on a hard floor might be too rough on your brick. We found that running these robots on a smooth carpet is the best of all options.**

ribbit: an NXT frog

Sometimes discoveries are made through simple observation. When I showed this robot design to a friend who is an engineer, his first comment was, "It's so simple, I never would have thought of it!"

This is the first jumping robot I designed. The stance is wide to make it look a little bit like a leaping frog—and to keep it from falling over.

Figure 3-1: An NXT frog

building ribbit

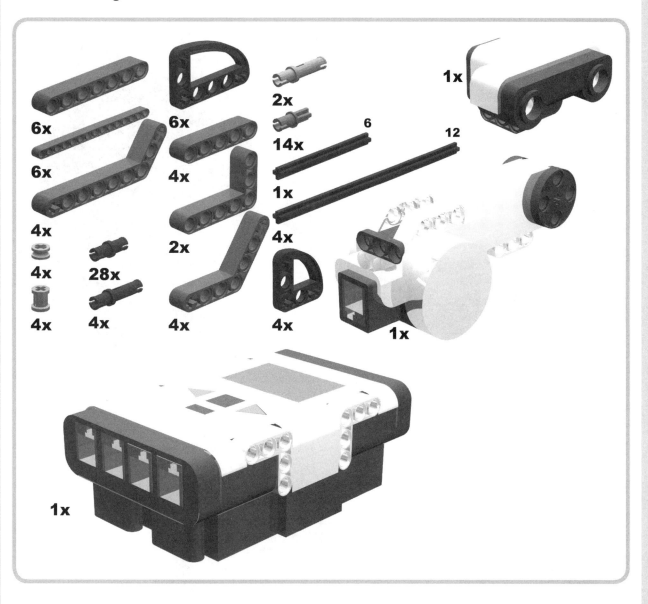

6x
6x
4x
4x
4x

6x
4x
2x
4x

2x
14x
1x
4x

28x
4x

4x
4x

6
12

1x
1x

1x

1x

* Non-LEGO parts required: Two rubber fingertips, available in office supply stores

In these steps, the screen on the brick is at the top.

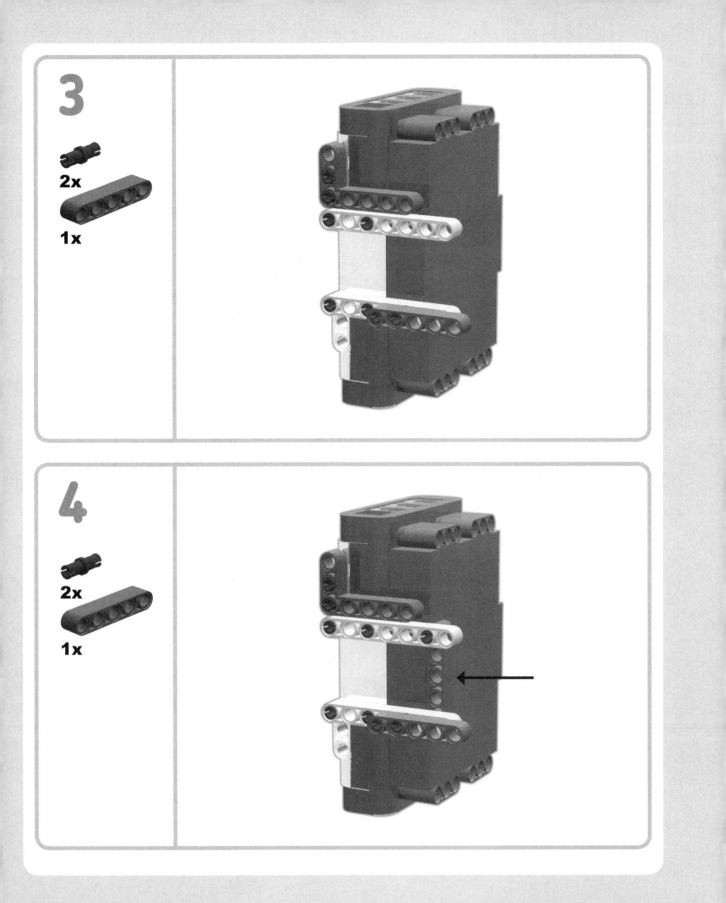

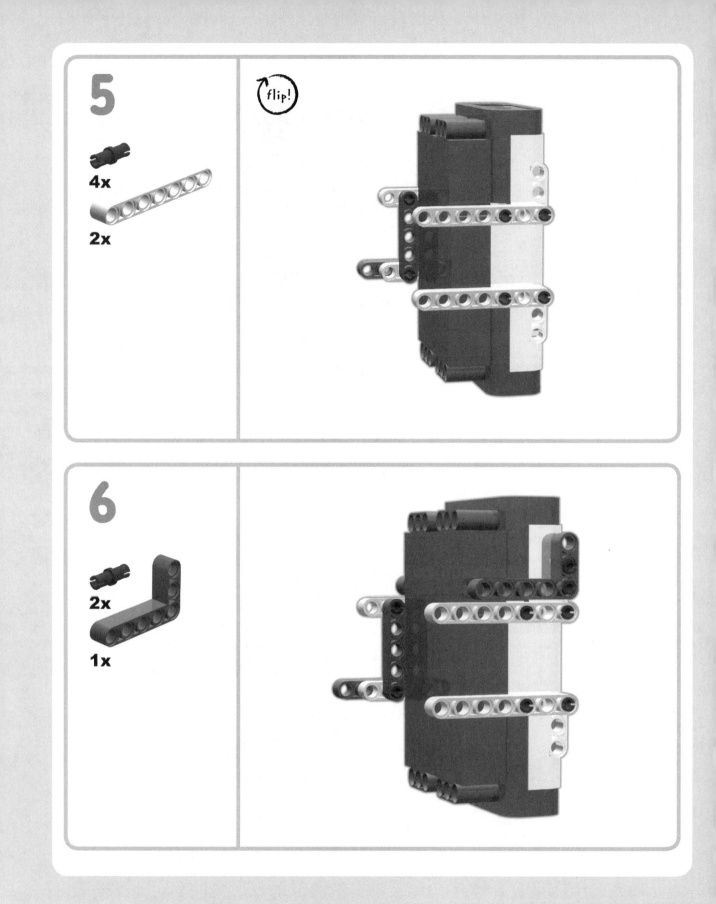

7

2x

1x

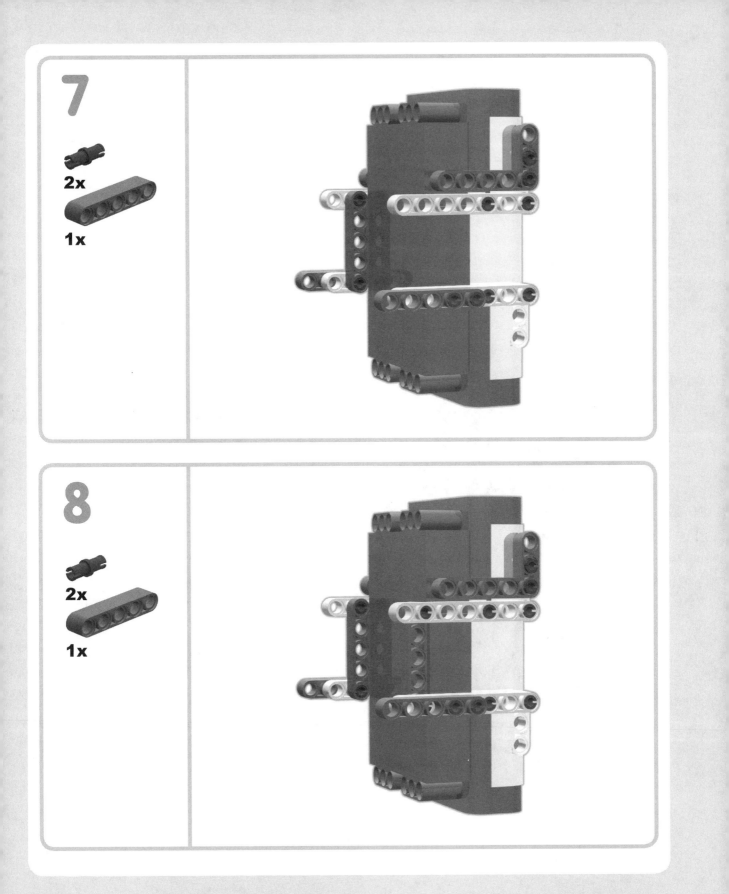

8

2x

1x

9

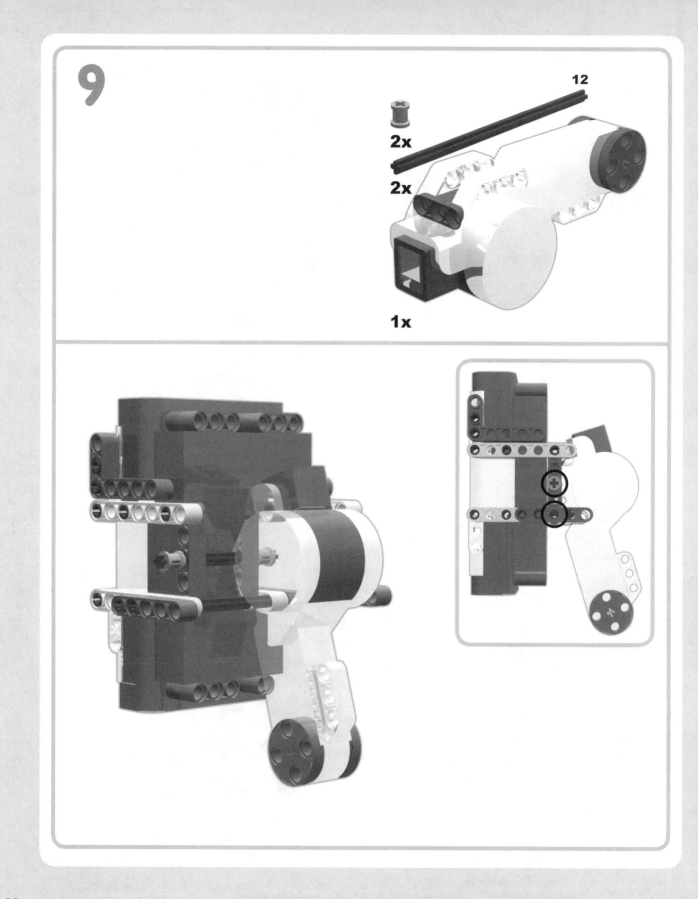

2x

2x

1x

12

10

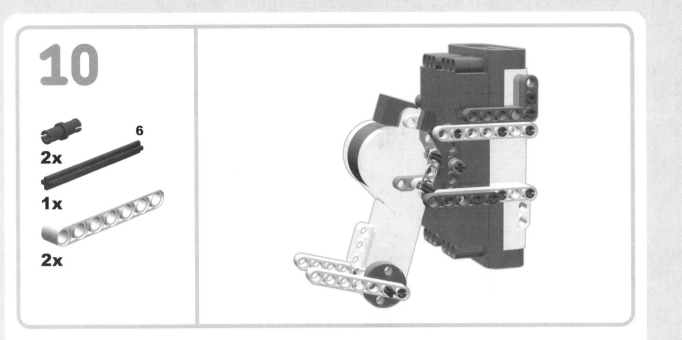

The 7-hole beam is held on with both an axle and a pin.

11

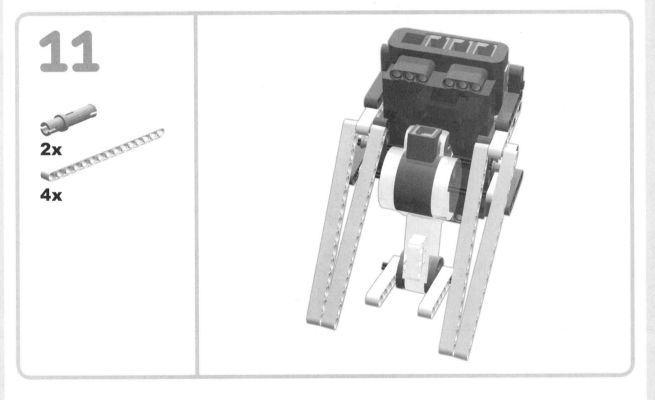

These four beams will hang loosely. Without these beams, the robot would just move up and down. At the end of these building instructions, I will show you how to make them work as you need them to.

12

3x

1x

2x

13

3x

1x

2x

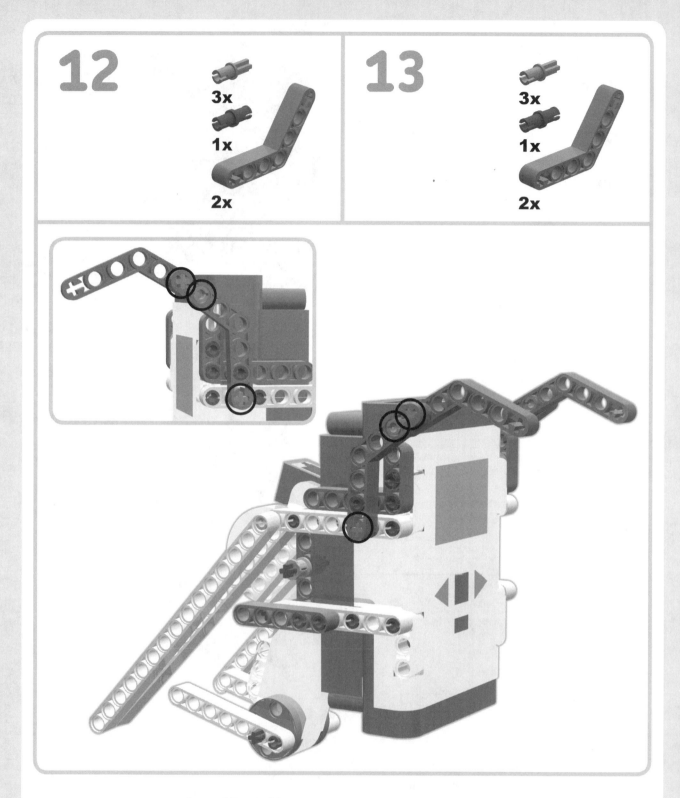

Steps 12 and 13 use the same parts on opposite sides. Use blue axle pins where marked.

14

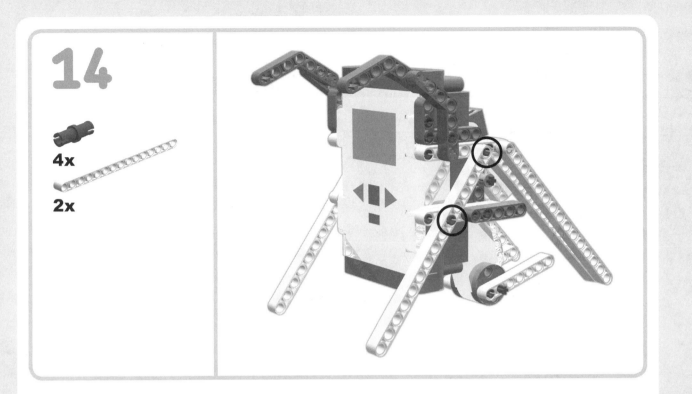

Put one beam and two pins on each side.

15

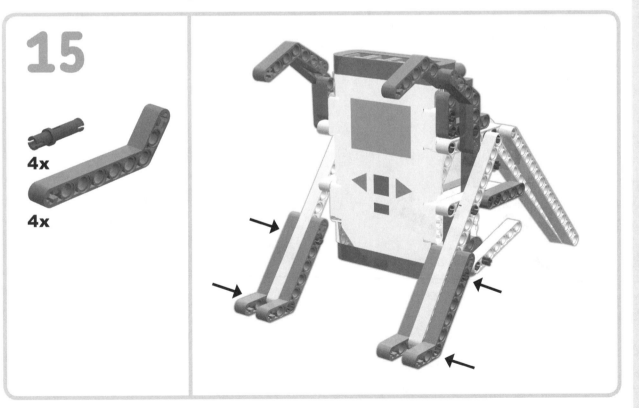

PART SUBSTITUTION SUGGESTION

The 15-hole beams in the front legs can be substituted with 13-hole beams and assembled as shown here.

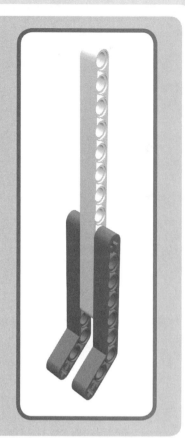

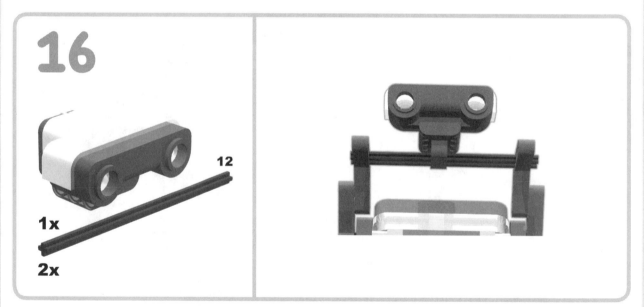

Put the axles in the third and fourth holes from the top end of the beam.
(If this is unclear to you, step 19 offers a side view.)

17

4x

2x

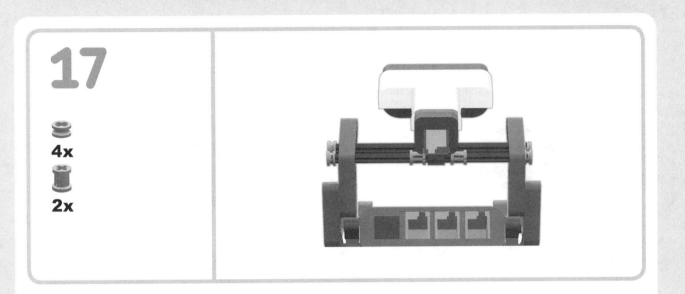

Steps 18 and 19 are meant to make your robot look more like a frog. They are not necessary for Ribbit to work correctly.

18

3x   **2x**

4x

four pins

These parts are placed on the inside of the frame (on the axles) near the head and on the outside on the legs. If you put them on the inside of the legs, they will get in the way.

19

3x **2x**

4x

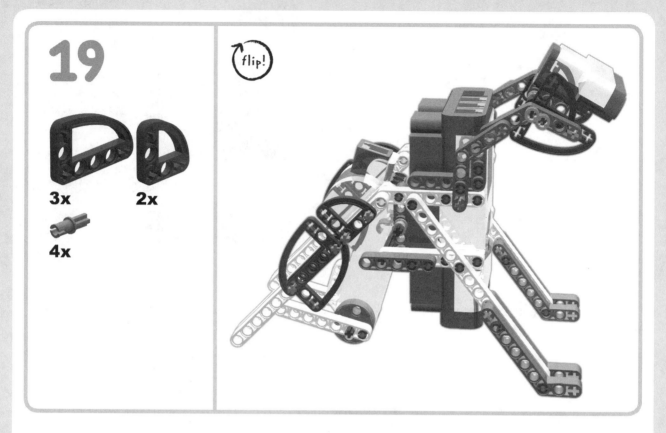

This robot can hop forward only if we keep it from moving backward. We do that by giving the legs some extra traction.

Capping each pair of back legs with a rubber fingertip (as shown in step 20) works very well. (These fingertips can be found in any office supply store.)

If the rubber fingertips are too loose, insert one of the smaller pins in your kit into the second hole from the bottom of one of the beams. This will hold the rubber fingertips securely.

I strongly recommend that this robot be run on a carpeted surface.

If your robot comes apart when hopping, try modifying the power or connecting the front legs to the body with long pins.

wiring connections

Using the shortest power cord you have, connect the motor to port A at the top. (This port gives the best power and is usually the best choice when using one port.) Connect the Ultrasonic Sensor (Ribbit's head) into port 4 at the bottom of the brick.

20

bunny: an NXT rabbit

Bunny is a rabbit that actually hops—without falling apart. It is a variation on the frog, designed to see how varying the angles and shapes would affect the hopping action. To me, it seems that the hopping is more delicate in the bunny than in the frog.

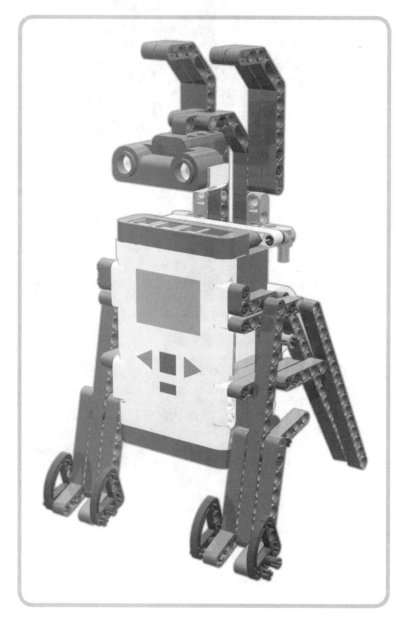

Figure 3-2: A hopping NXT Bunny

building bunny

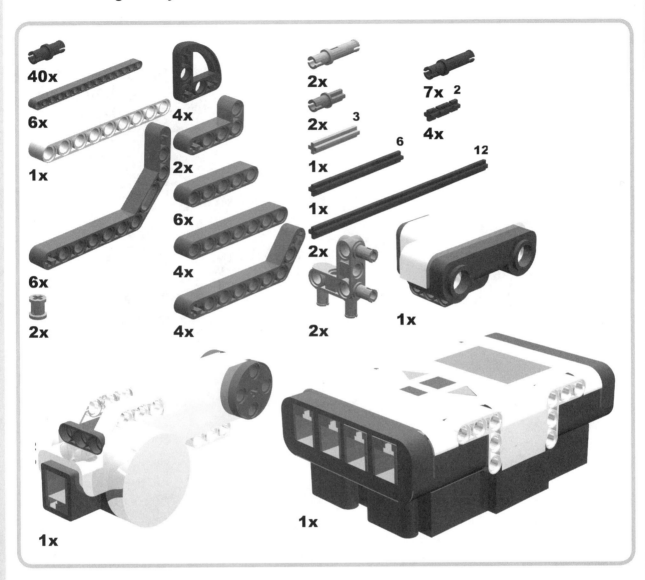

40x

6x

1x

6x

2x

4x

2x

6x

4x

4x

2x   3

1x   6

1x

2x

2x

7x   2

4x

12

1x

1x

1x

* Non-LEGO parts required: Two rubber fingertips, available in office supply stores

In these steps, the screen on the brick is at the top.

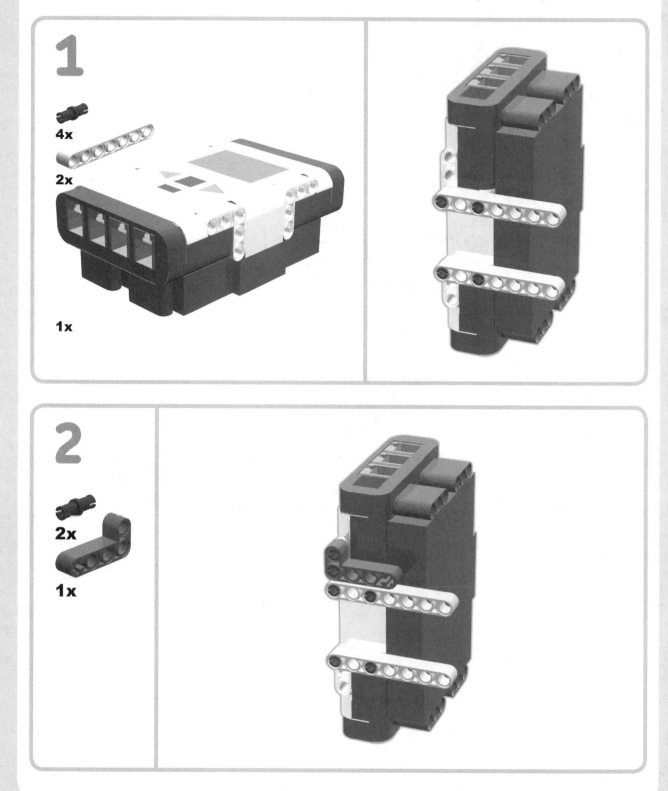

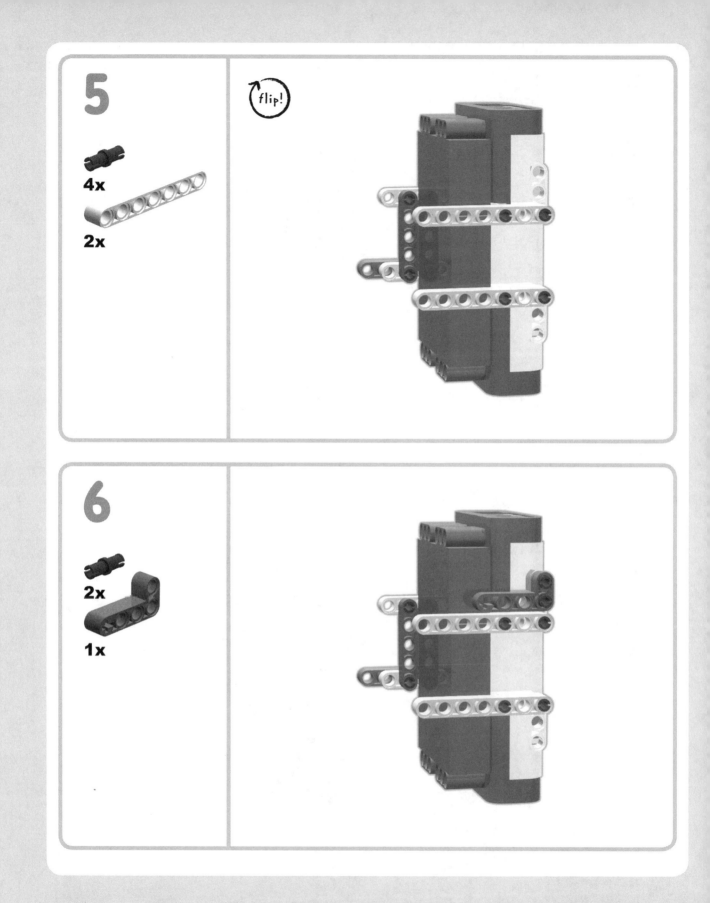

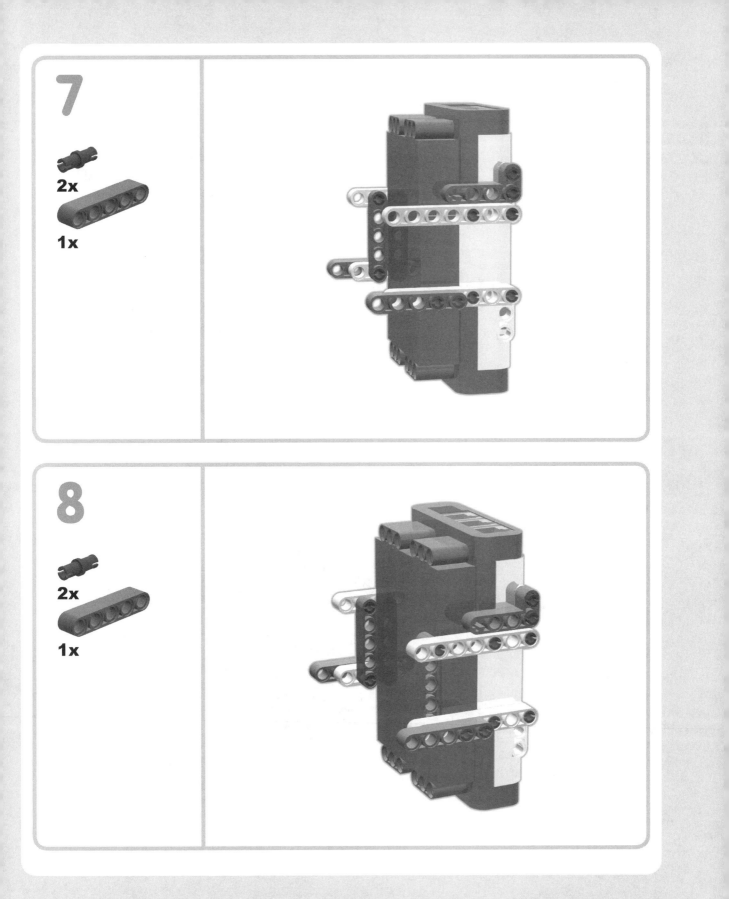

11

2x

4x

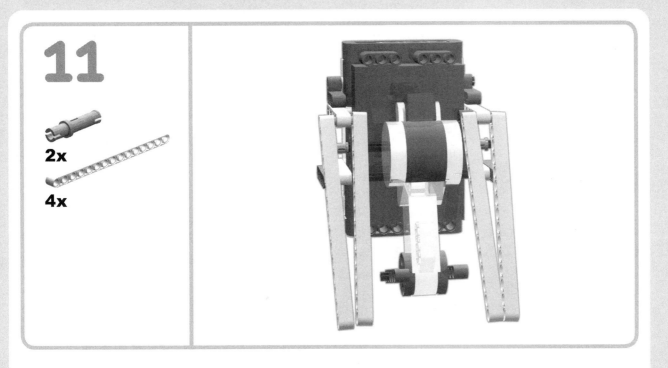

These four beams will hang loosely. Without these beams, the robot would just move up and down. At the end of these building instructions, I will show you how we make them work as we want them to.

12

4x

2x

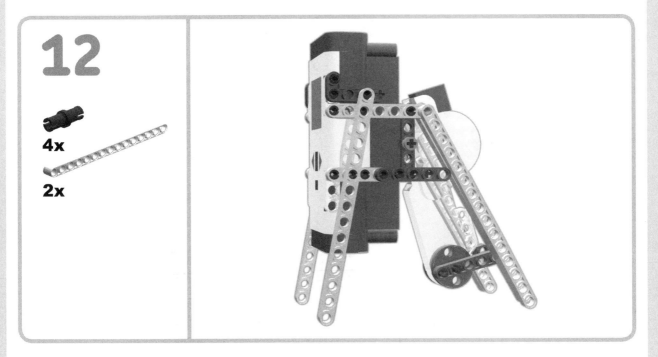

Add one beam on each side.

PART SUBSTITUTION SUGGESTION

The 15-hole beams in the front legs can be substituted with 13-hole beams and assembled as shown here.

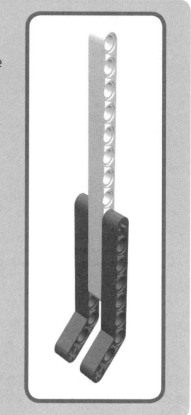

13

4x

4x

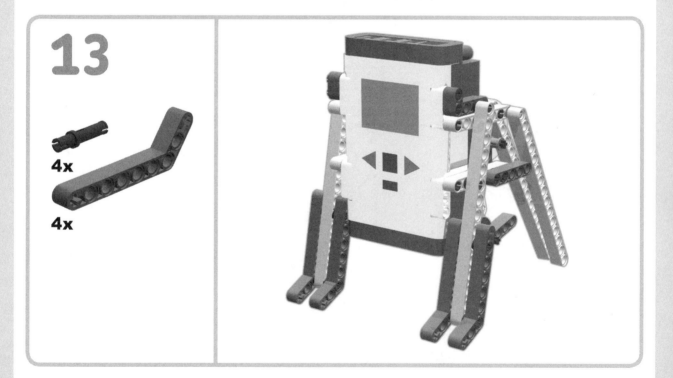

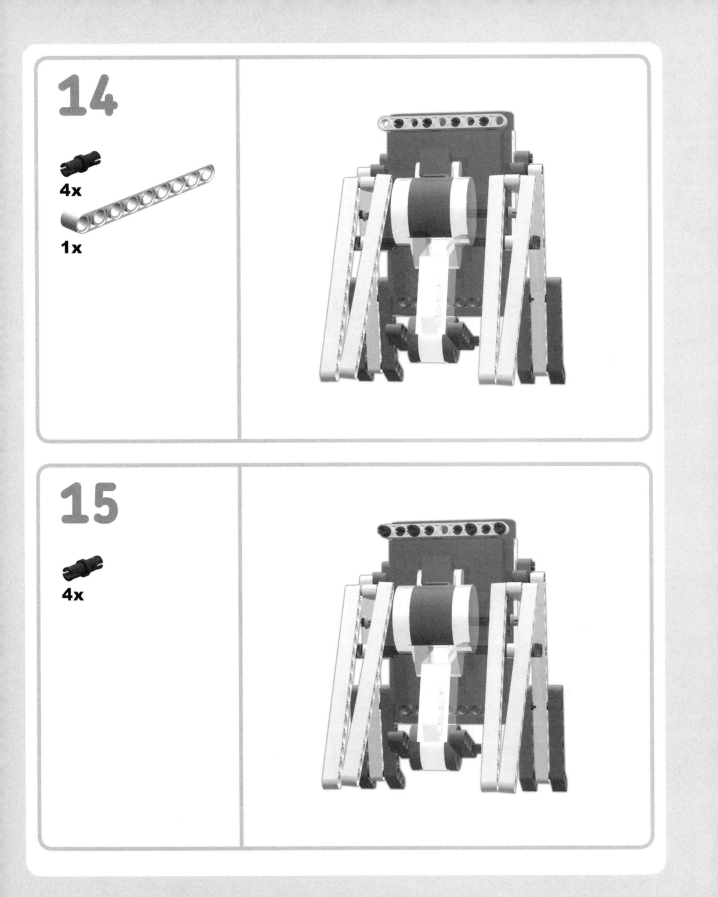

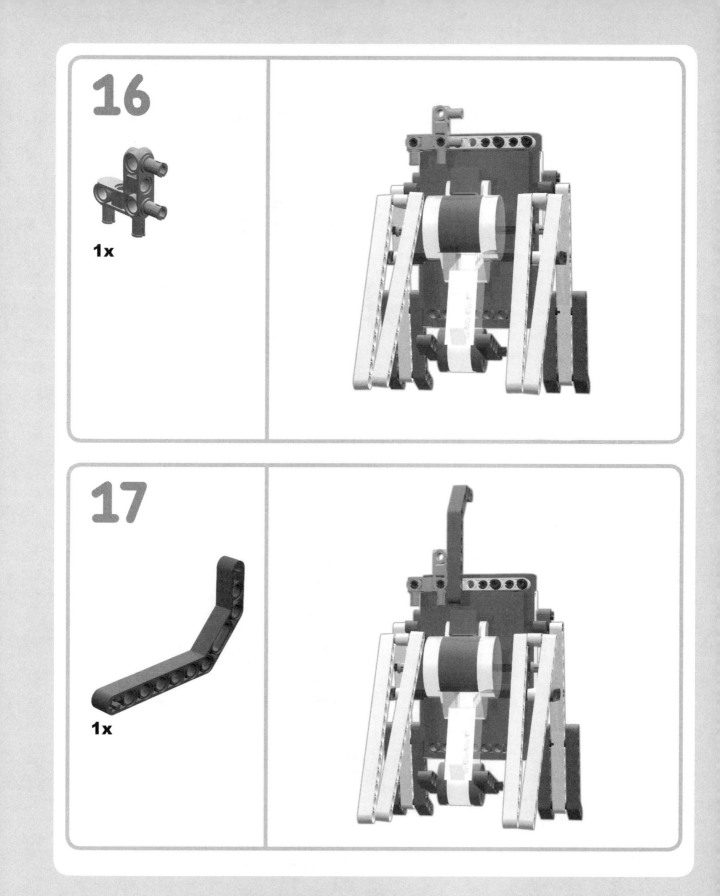

16

1x

17

1x

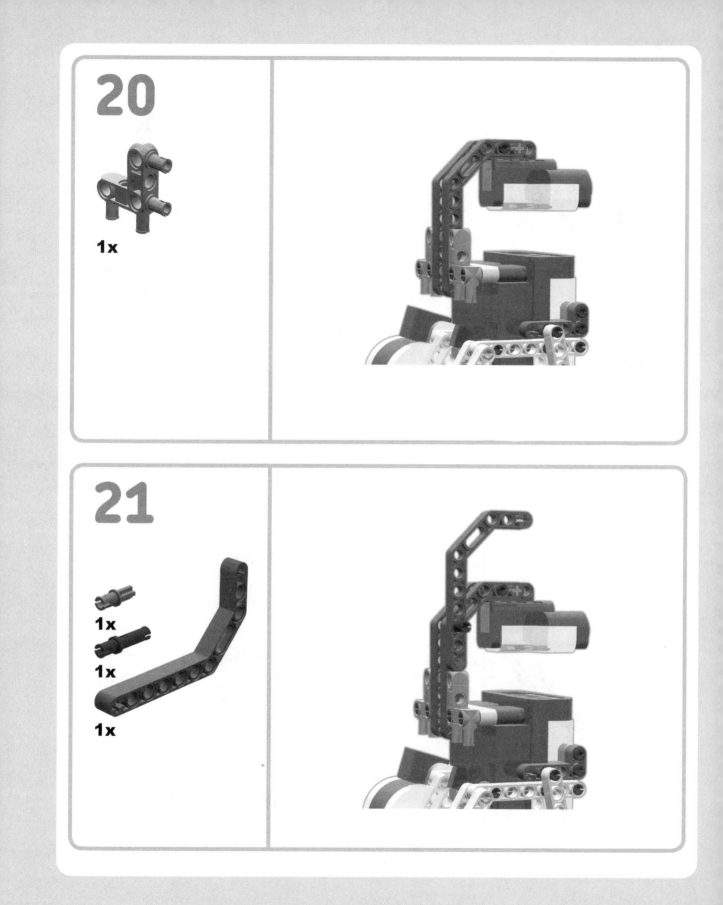

20

1x

21

1x

1x

1x

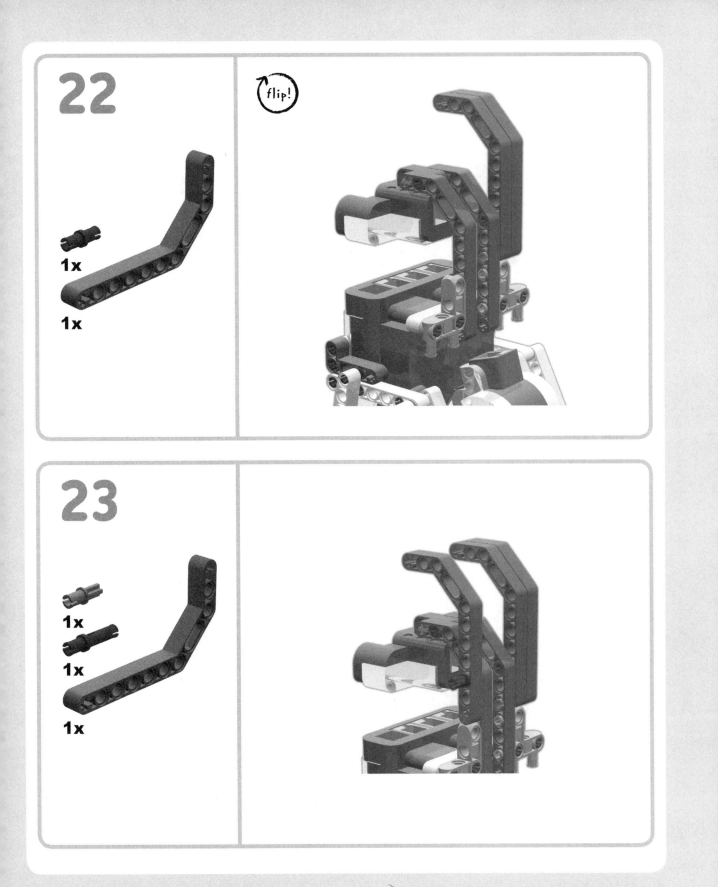

NOTE The LEGO pieces in step 25 are just decorative. Your robot will work fine without them. If you don't have them, you might just have another part that would contribute the shape of a rabbit's foot.

If your robot comes apart when hopping, try modifying the power or connecting the front legs to the body with long pins.

For Bunny to move forward, we must keep it from sliding backward. We'll do that by capping the pairs of loosely-hanging back legs with rubber fingertips. (These fingertips can be found in any office supply store.) The rubber fingertips are used on Bunny's rear legs, just as they were on Ribbit (shown in Figure 3-3).

If a rubber fingertip is too loose, insert one of the smaller pins from your kit into the second hole from the bottom of one of the beams. Doing this will hold the rubber fingertips securely.

I recommend that you run Bunny on carpet to minimize the stress on your brick.

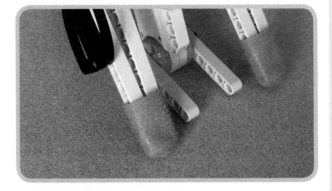

Figure 3-3: Rubber fingertips go on Bunny's back legs in the same way as shown here on Ribbit.

wiring connections

Using the shortest power cord you have, attach the motor to port A at the top. (This port gives the best power and is usually chosen when using one port.) Connect the Ultrasonic Sensor into port 4 at the bottom of the brick.

programming ribbit and bunny

Hopping Program #1 shows you how to make your robots hop in a random pattern. This hopping would be more similar to a real animal.

Hopping Program #2 gives you more control over how many hops your robot makes—and in what pattern. This will be helpful when you want your robots to follow specific directions.

Both programs can be used with either robot. It's up to you to choose.

hopping program #1: random hopping

This program produces a random pattern of hops.

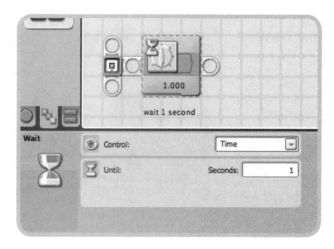

Figure 3-4: We begin with a Wait block. Configure it as shown. This pause allows you time to pull your hand away before the robot moves. (Make it any length of time you like.)

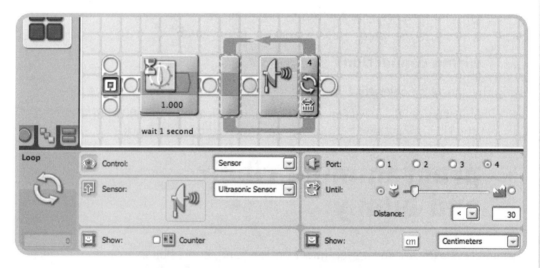

Figure 3-5: We don't want our robot to run into walls, so let's insert a loop configured with an Ultrasonic Sensor. Set the Distance to <30 centimeters. (Remember to change it to centimeters before you type your distance.)

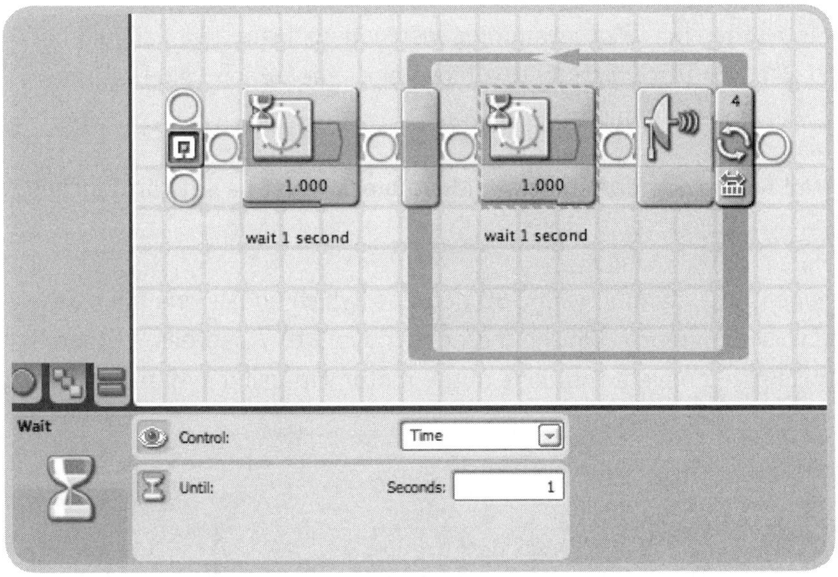

Figure 3-6: Placing a Wait block inside the loop doubles the time we have to get away after pressing the Run button. It also produces a 1-second pause between each series of jumps.

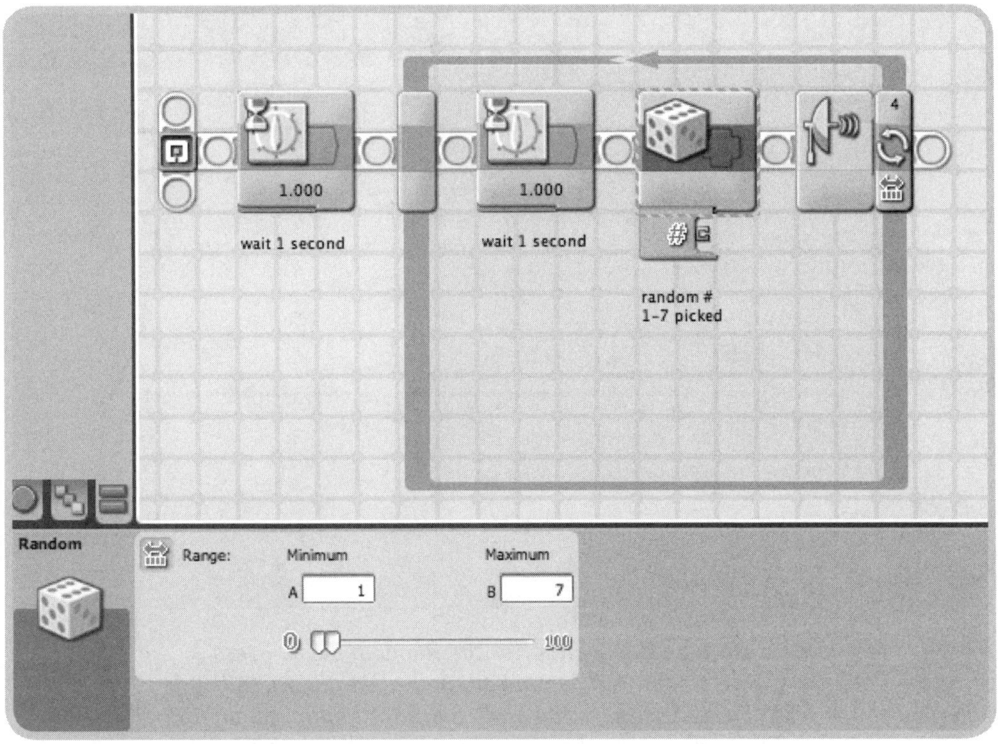

Figure 3-7: Find the Random block in the Complete palette and place that inside the loop. Configure it for a minimum of 1 and a maximum of 7. This means my robot will not hop more that 7 times consecutively. I chose that number because I want the robot to check often for solid objects ahead.

According to all the NXT-G programming instructions I can find, I should be able to connect my Random block directly to a Move block, but when I did that my robot didn't hop; it was more of a hiccup.

NOTE When paired with a **Random block**, a **Move block** must be set for **degrees**, not rotations.

However, when I set my motor for 360 degrees (which equals one full rotation), the Random block apparently told the motor to move 1 to 7 degrees, instead of 1 to 7 rotations. To get complete rotations of the motor, I'm using a Math block.

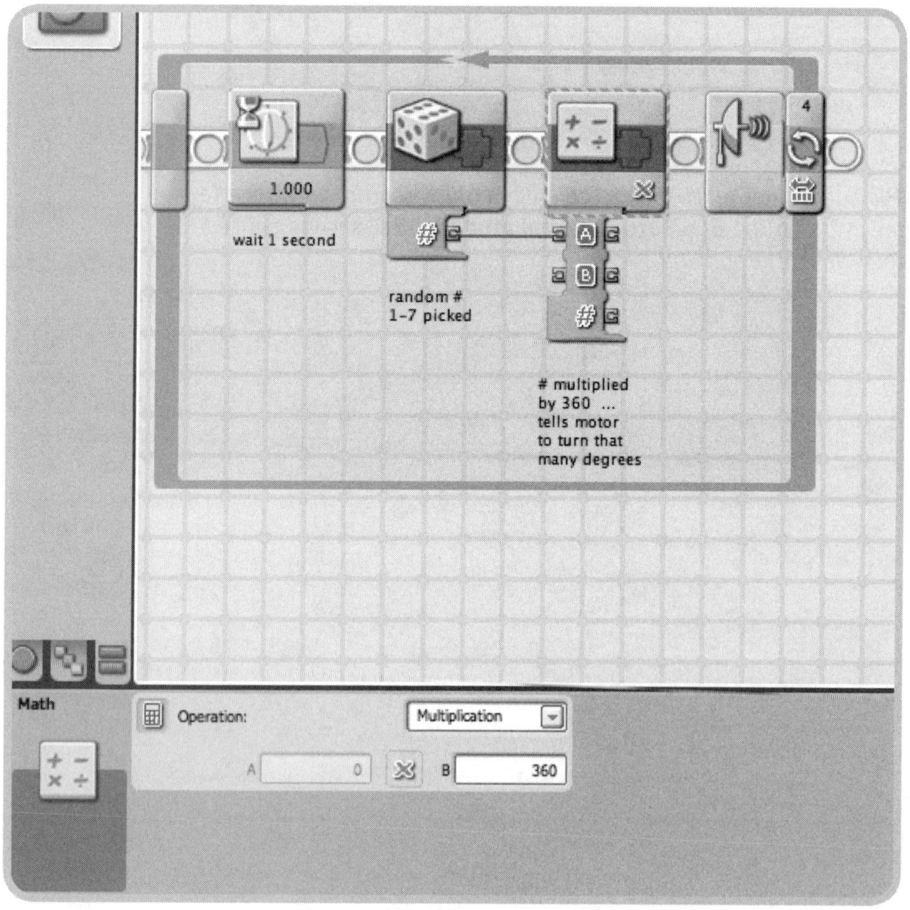

Figure 3-8: Place a Math block on the beam after the Random block. Press the SHIFT key and drag a wire from the # symbol on the Random block to the A data-in plug on the Math block. Configure the Math block for Multiplication and type 360 in the B box. The Math block takes any number sent from the Random block, puts it in the A box, and multiplies it by 360.

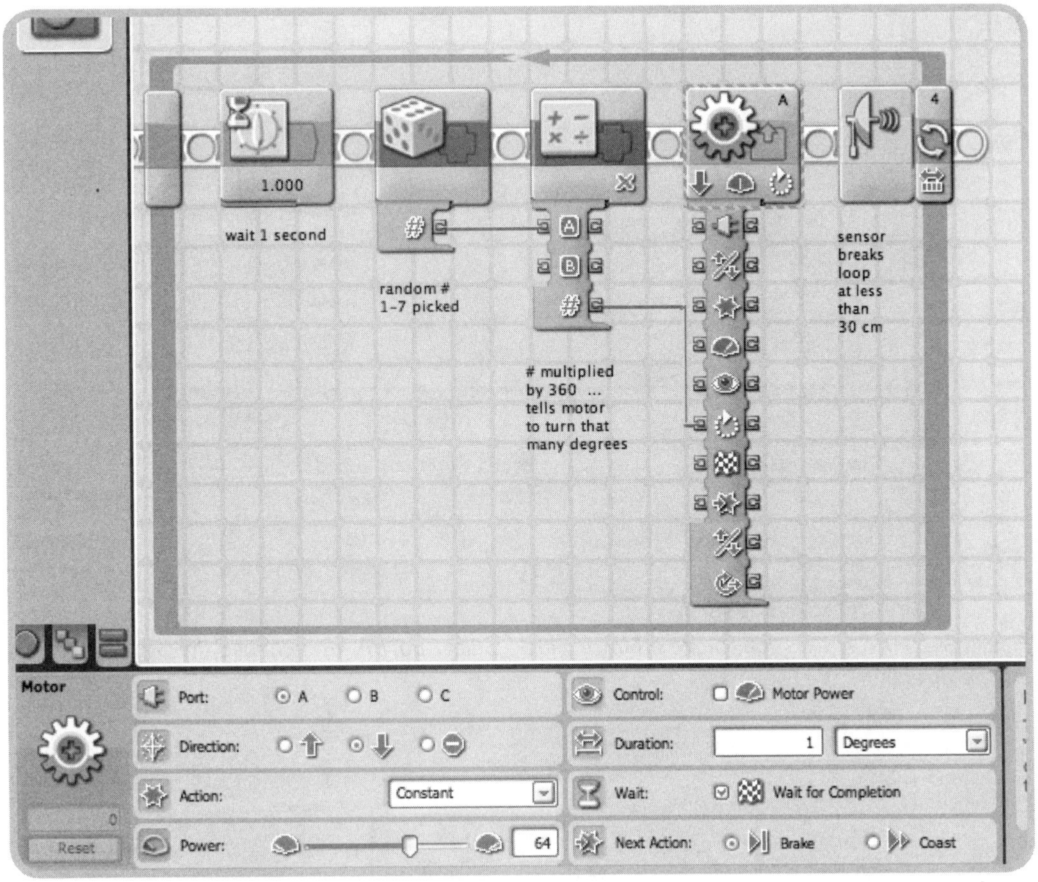

Figure 3-9: Follow the Math block with a Motor block. Configure it as shown. Press SHIFT and drag a wire from the # plug on the Math block hub to the Duration input plug on the Motor block hub. The Math block will tell motor A how many times to turn using degrees. (If you don't see the hub shown in this figure, click in the lower-left corner of the block, and the hub should drop down.)

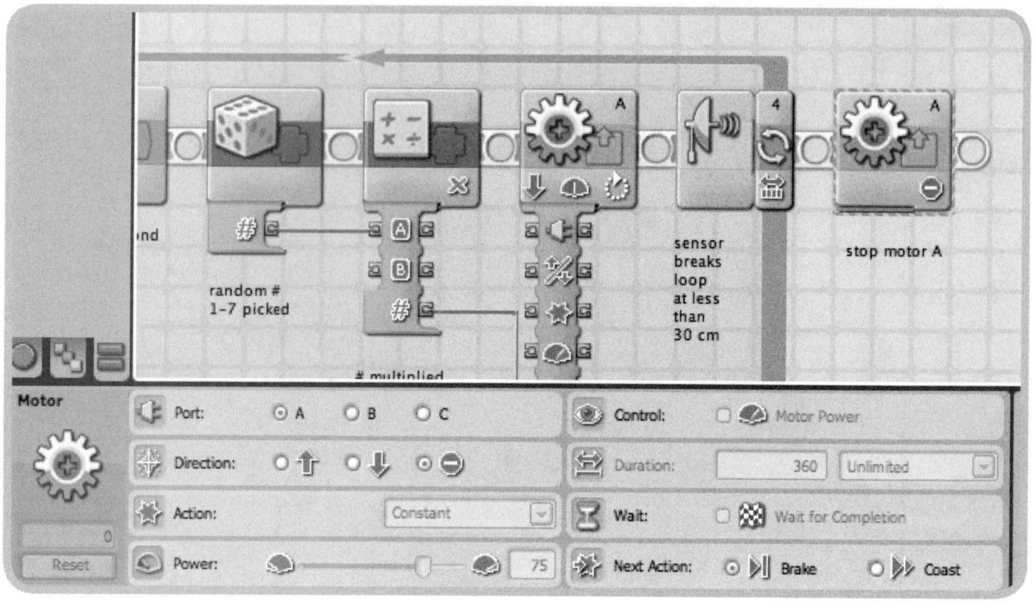

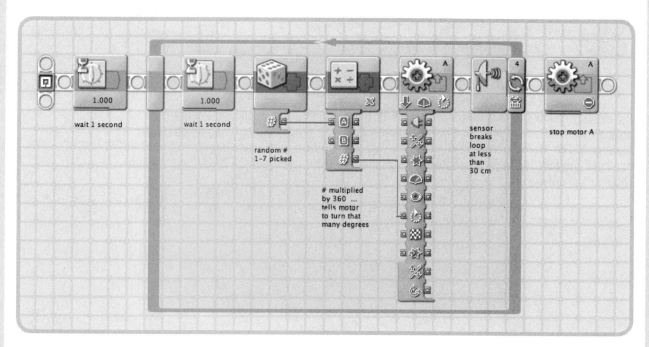

Figure 3-10: After the loop, place a Motor block. Select port A, and choose the Stop button for the direction.

Figure 3-11: The complete program

Now you're ready to save the program and download it to your NXT brick.

NOTE As with all the robots, you will need to modify the power settings for your specific conditions—battery strength and hopping surface, in particular.

hopping program #2: create your own pattern

The first thing we'll do is create a series of nearly identical My Blocks.

my blocks: 1Hop, 2Hops, 3Hops, 4Hops, 5Hops, 6Hops . . .

These blocks will all be in the form of switches. Each will decide whether the robot will stop or keep going.

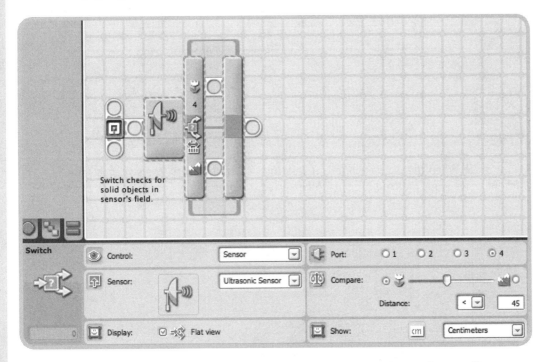

Figure 3-12: Here is our Switch block, configured as an Ultrasonic Sensor (in port 4) to scan for a distance of less than 45 centimeters.

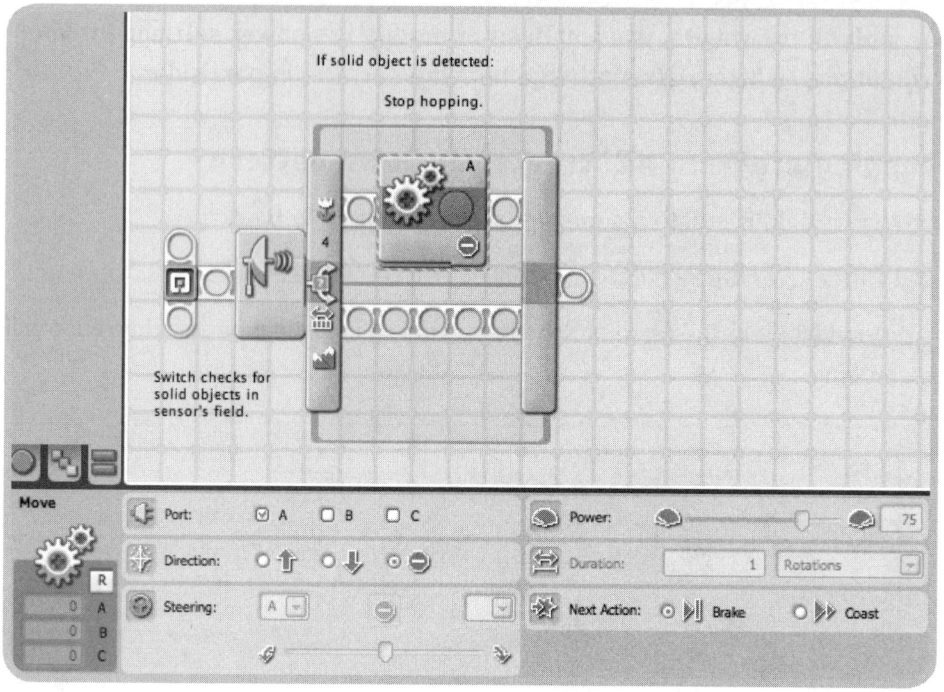

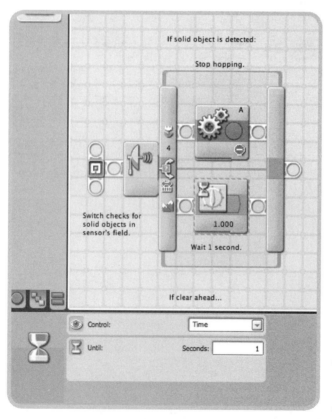

Figure 3-13: When the distance ahead is less than 45 centimeters, the switch is True. If it's True, we want the frog to stop, so I inserted a Motor block. Configure it as shown.

Figure 3-14: If the answer is False (the robot is not closer than 45 centimeters to a solid object), I want the robot to stop for 1 second and then continue to the next block. So, my first block on the bottom line is a Wait block, configured to wait 1 second.

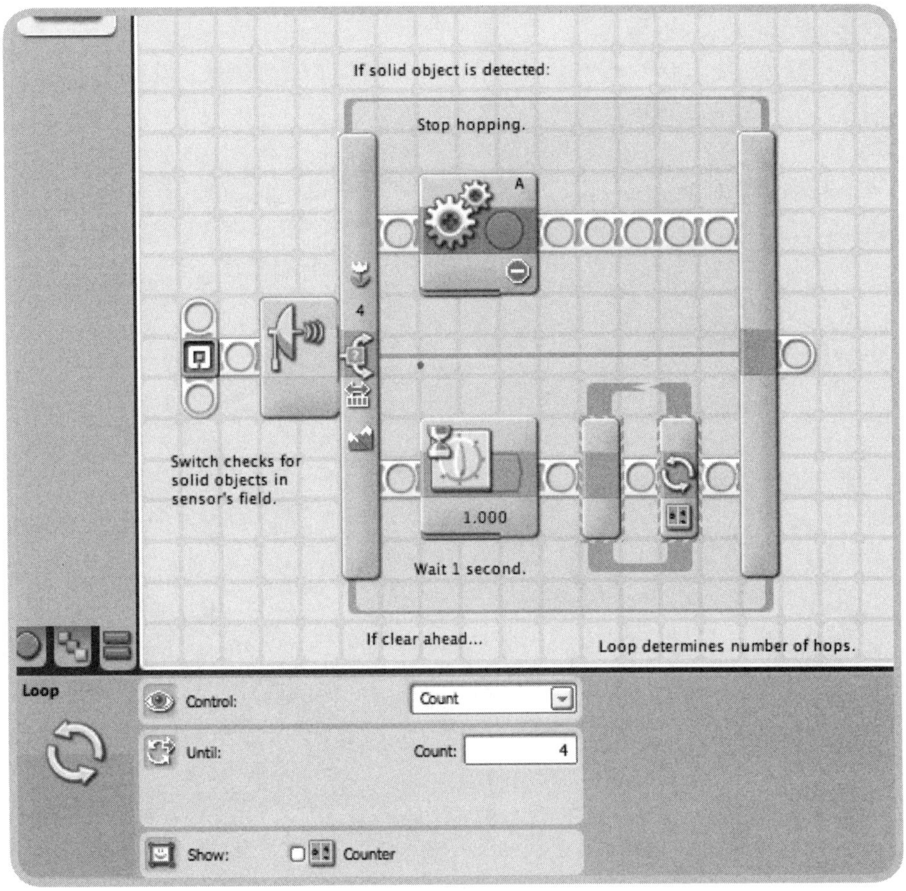

If solid object is detected:

Stop hopping.

Switch checks for solid objects in sensor's field.

Wait 1 second.

If clear ahead...

Loop determines number of hops.

Loop

Control: Count

Until: Count: 4

Show: Counter

Figure 3-15: This next block is a loop, configured to count. Type the number 4, as shown in the figure. This loop will now repeat 4 times.

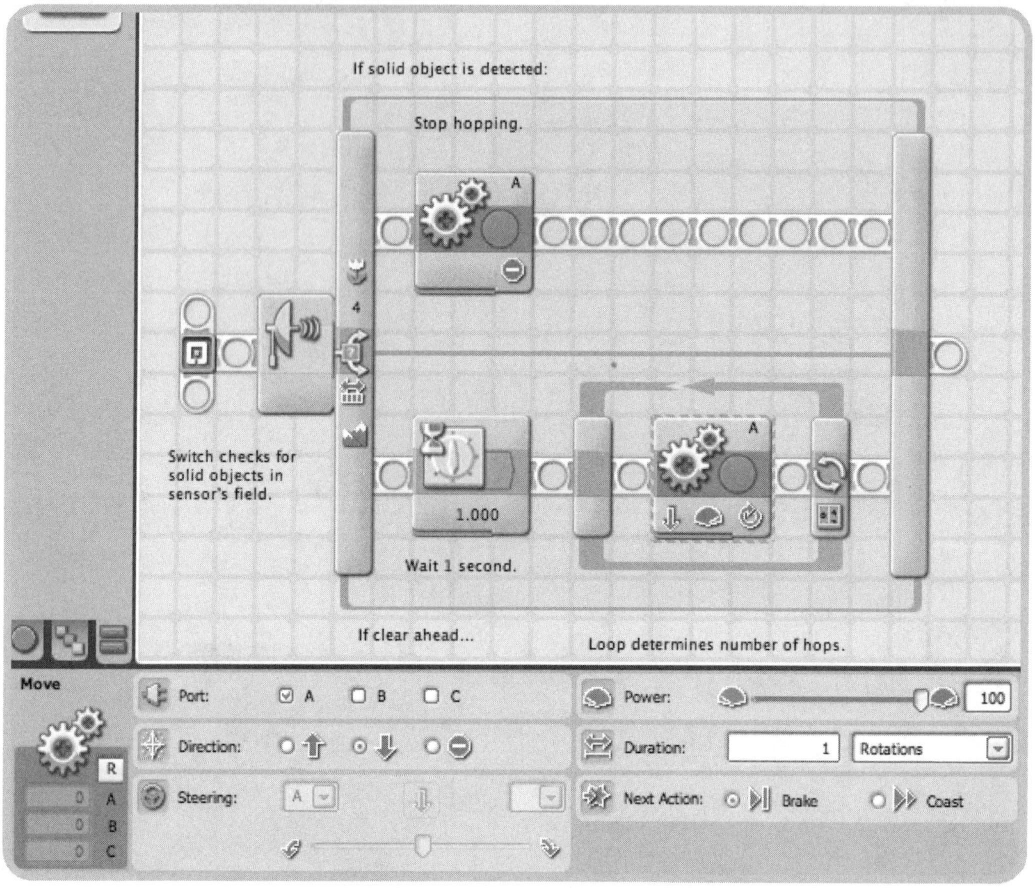

If solid object is detected:

Stop hopping.

Switch checks for solid objects in sensor's field.

Wait 1 second.

If clear ahead...

Loop determines number of hops.

Figure 3-16: Inside the loop, we'll place a Motor block configured to make motor A turn one rotation backward (which will move your hopper forward).

The power setting for this robot's motor very much depends on how strong the charge is in your battery. Start low, say 50, and then increase. (The last time I built this robot, I set the power at 100 because my batteries were very weak.)

Select all the blocks with your mouse and save them as a My Block named *4Hops*.

Now repeat the process you used to create *4Hops*, and create *1Hop*, *2Hops*, *3Hops*, *5Hops*, and *6Hops*. Each of them will be the same, except for the count number on the inner loop. Make the count loop 1 for *1Hop*, 2 for *2Hops*, 3 for *3Hops*, 5 for *5Hops*, and 6 for *6Hops*.

main program

For the main program, you get to decide what pattern of hops your robot will perform.

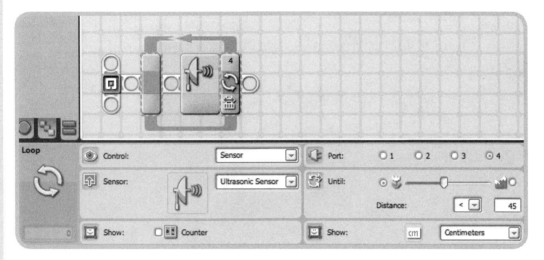

Figure 3-17: We'll begin by placing a loop on the beam and configuring it as an Ultrasonic Sensor. Use the settings shown here.

Now is the time to select from your assortment of new My Blocks and create your own series of hops (and pauses). This pattern of hops will be repeated until the robot senses a solid object in front of it (which makes it easy to stop the robot by putting your hand in front of the sensor).

series of my blocks within a loop configured to break with approach of object

| 6 hops | 2 hop | 4 hops | 1 hop | 3 hops | 5 hops |

Figure 3-18: My pattern of hopping blocks

When you have completed the hopping pattern for your robot, save the program and download it to the NXT brick.

sandy: an NXT camel

Using the tires and axles for feet and legs gives the impression of a camel's long, knobby legs and big feet. I hope it will also inspire you to think creatively about which parts you will choose when you design your own robots.

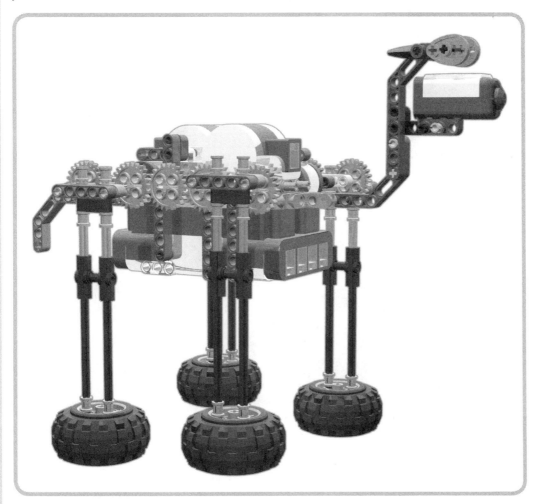

Figure 4-1: An NXT camel

building sandy

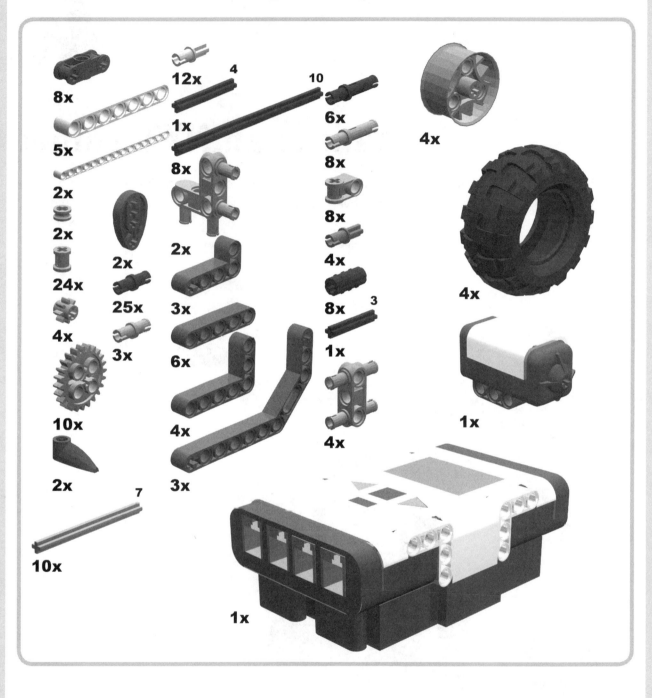

8x

5x

2x

2x

24x

4x

10x

2x

12x

1x

8x

2x

2x

25x

3x

3x

6x

4x

3x

4

10

6x

8x

8x

4x

8x   3

1x

4x

10x   7

4x

4x

4x

1x

1x

legs (build four)

If you build all four of these legs at once, it will help you make them all the same.

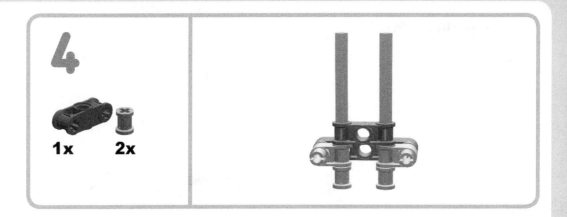

Push the parts on the axles all the way down to the end of the axles, as shown in step 4.

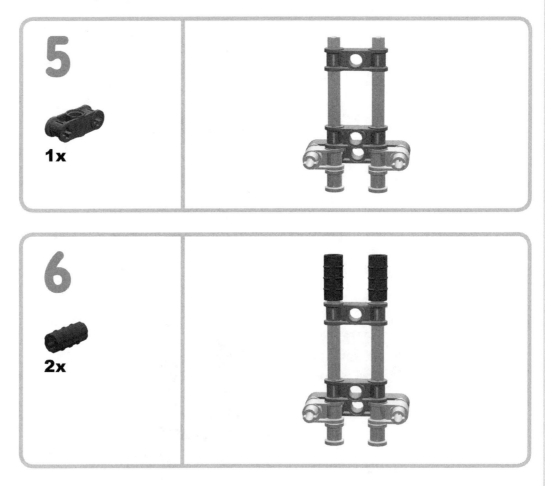

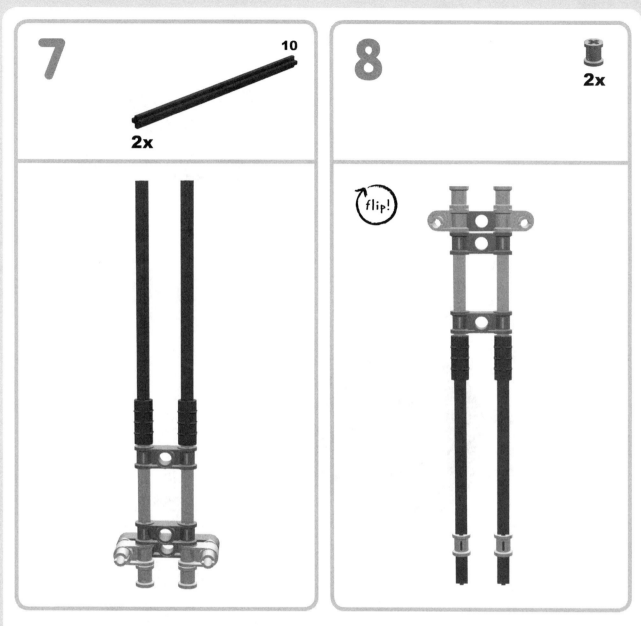

7

10

2x

Push these two #10 axles down firmly into the open ends of the axle connectors, as shown.

8

2x

flip!

Push these bushings onto the two open ends of the axles; then turn the whole structure over as shown.

Assemble the wheels and tires. Push the lower open end of the axles down into two of the holes in the wheel hub. Under the wheel, put bushings over the open ends of the axles, as shown here.

Set the leg down as shown and push down on the entire leg. Then push down on the bushings that are just above the tire. This will help ensure that the legs are all the same length and that the structure is secure.

full assembly

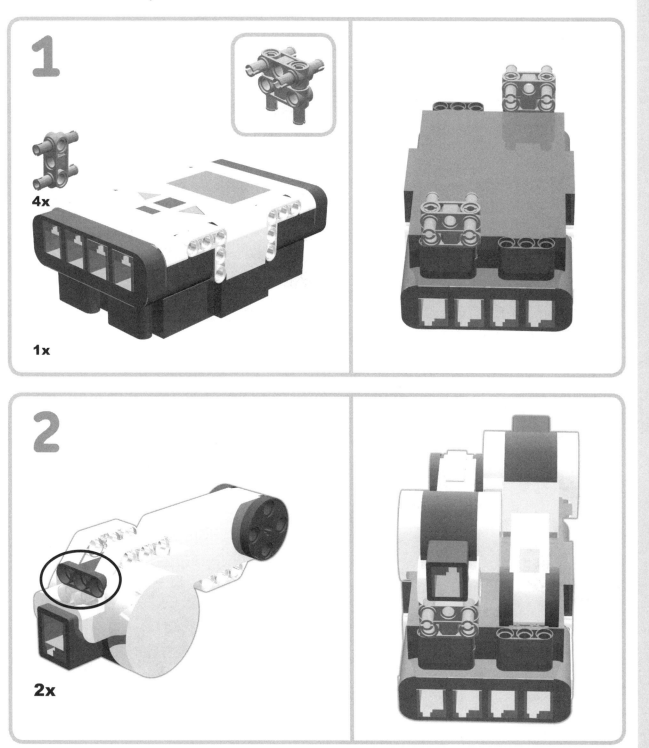

The pins on the connectors slip nicely into these holes.

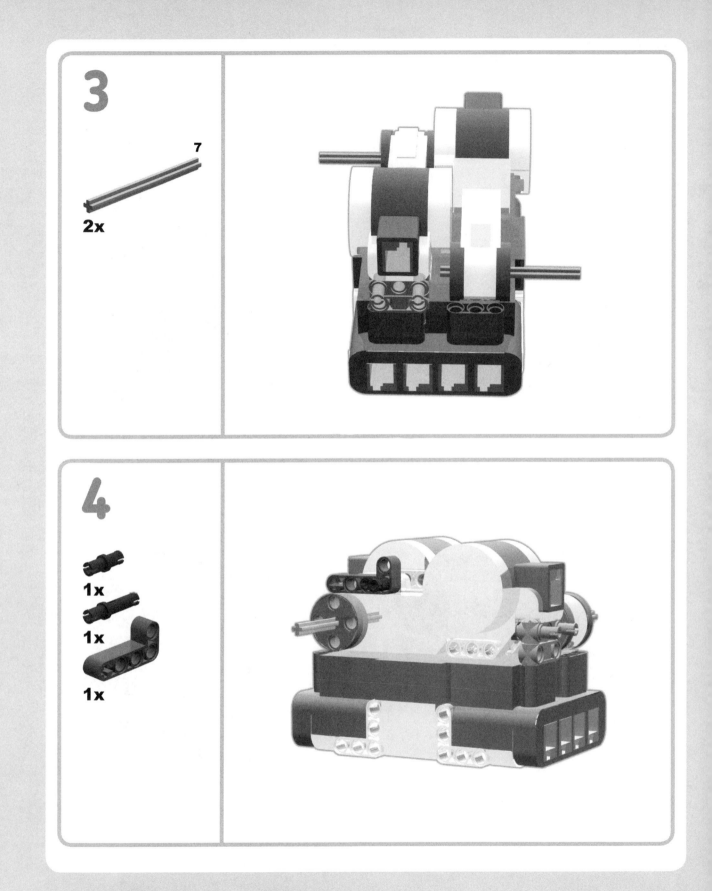

3

7

2x

4

1x

1x

1x

5

1x

1x

6

1x

1x

Thread the axle through the second hole in the 5-hole beam.

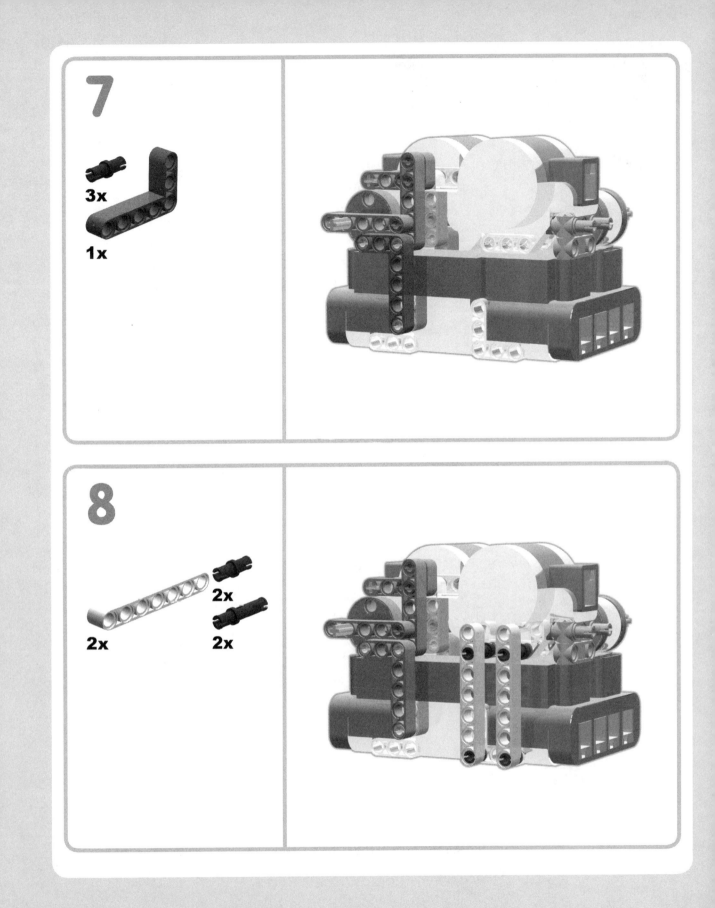

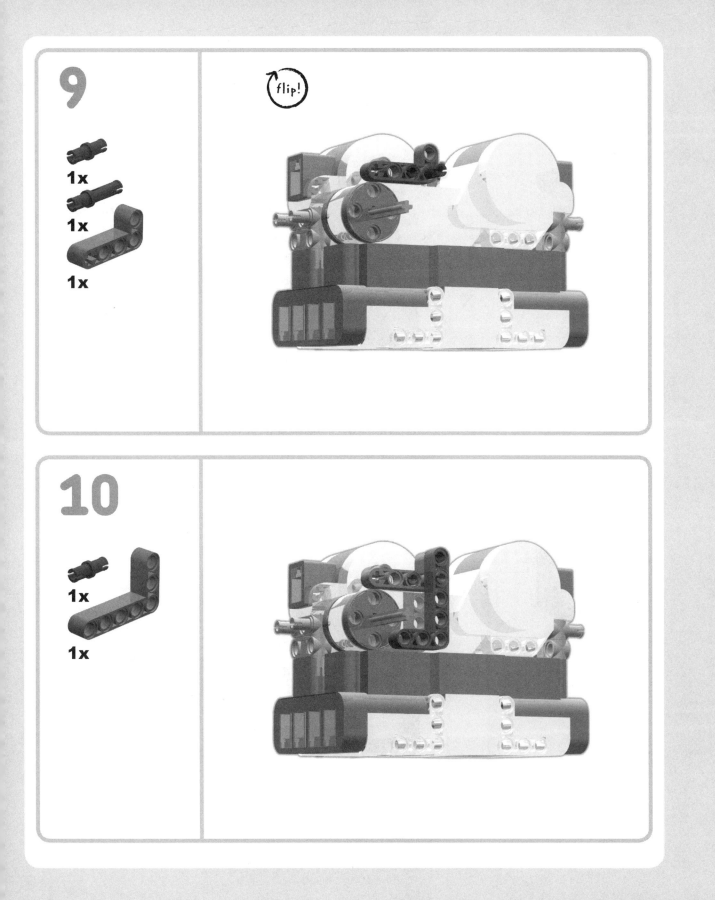

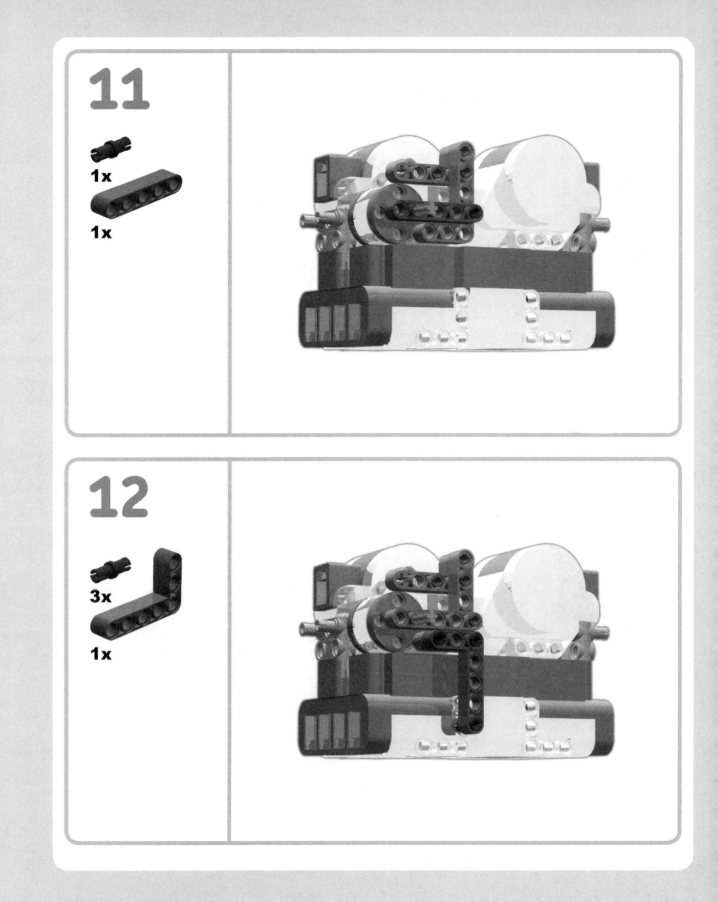

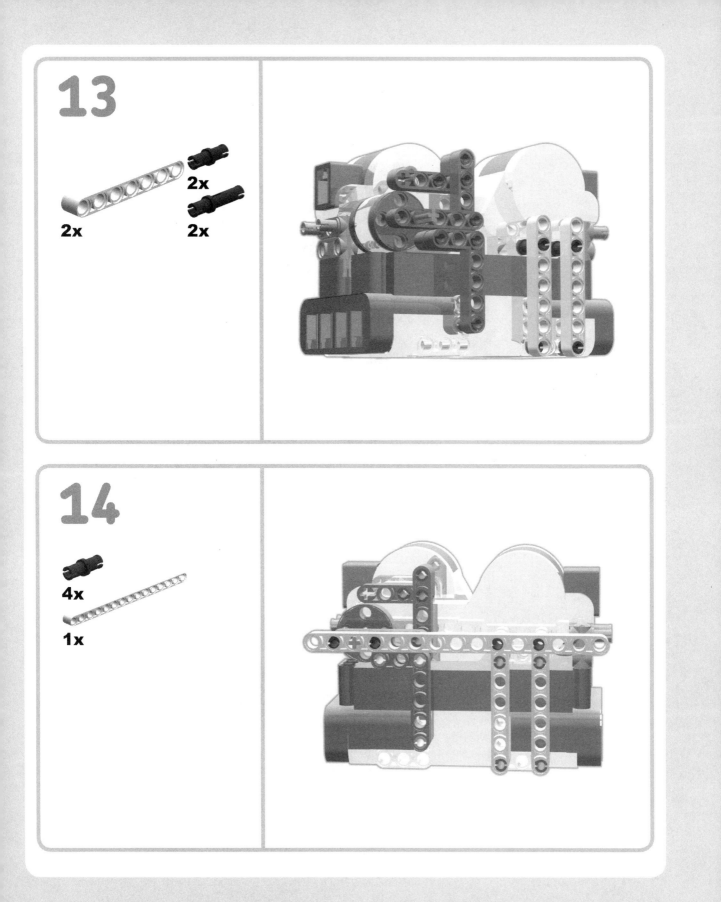

15

4x

1x

flip!

16

6x

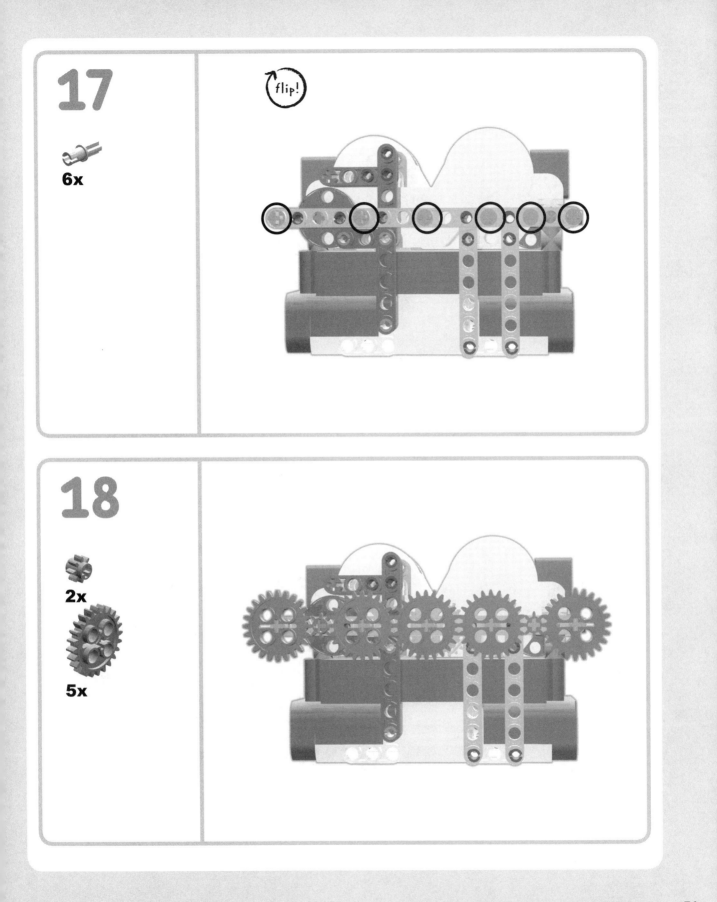

19

2x

5x

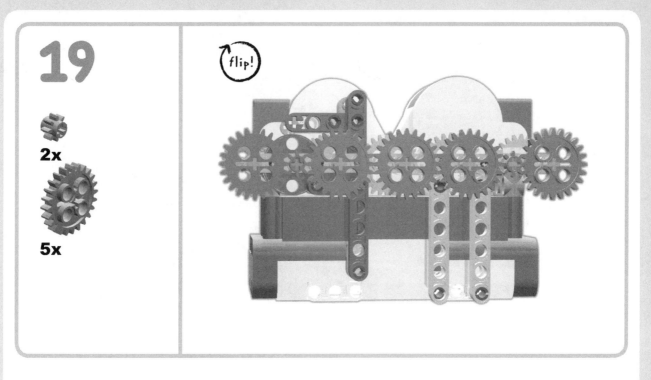

The holes on the gears should line up exactly in the position shown in step 19. If they move out of alignment as you build the robot, wait until it is time to attach the legs to rearrange them.

20

1x

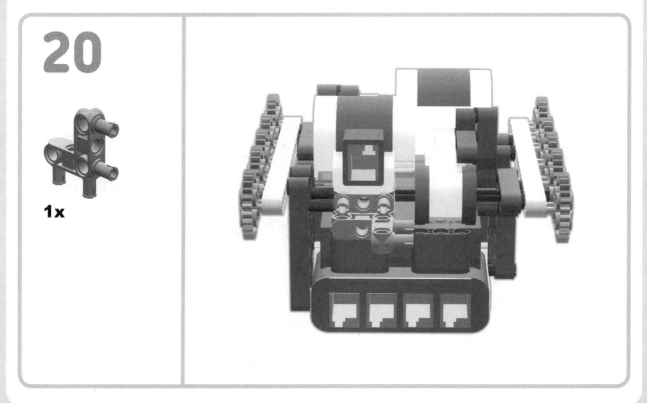

21

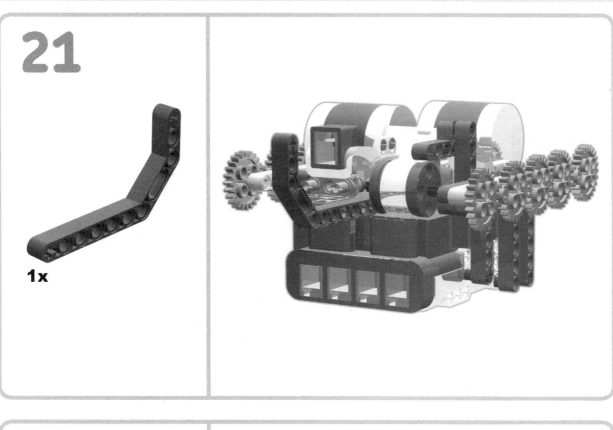

1x

22

2x

1x

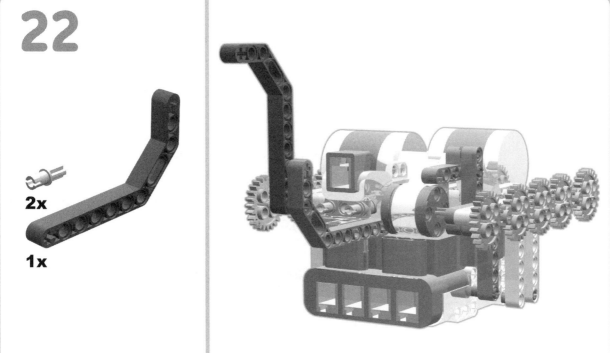

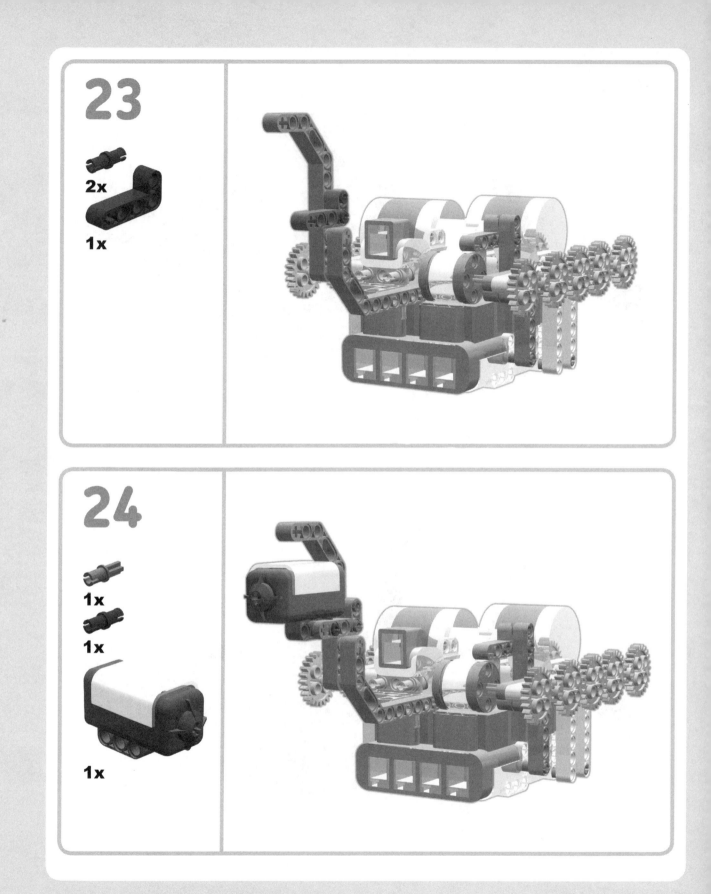

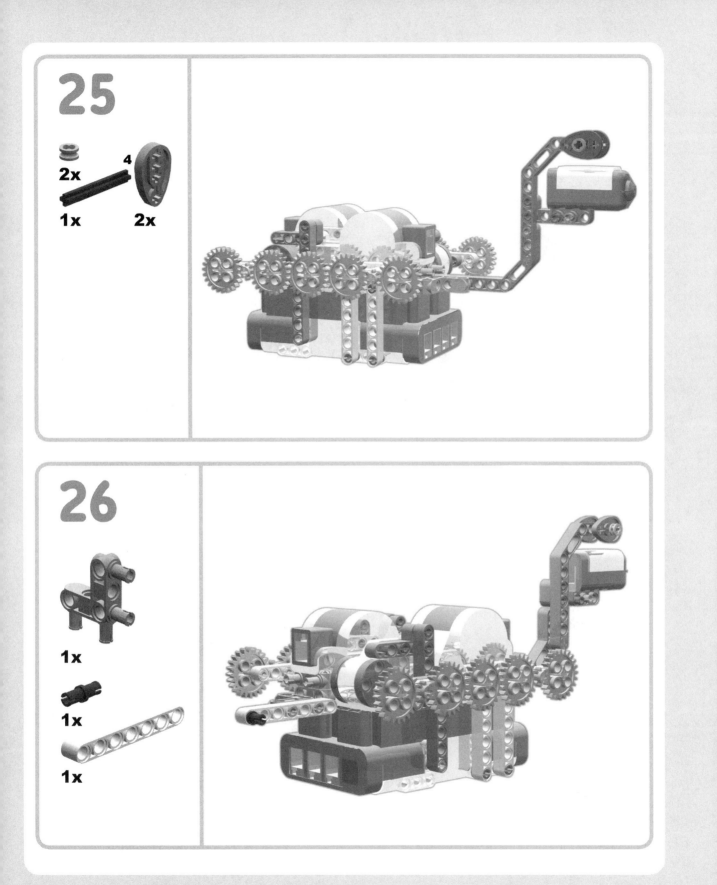

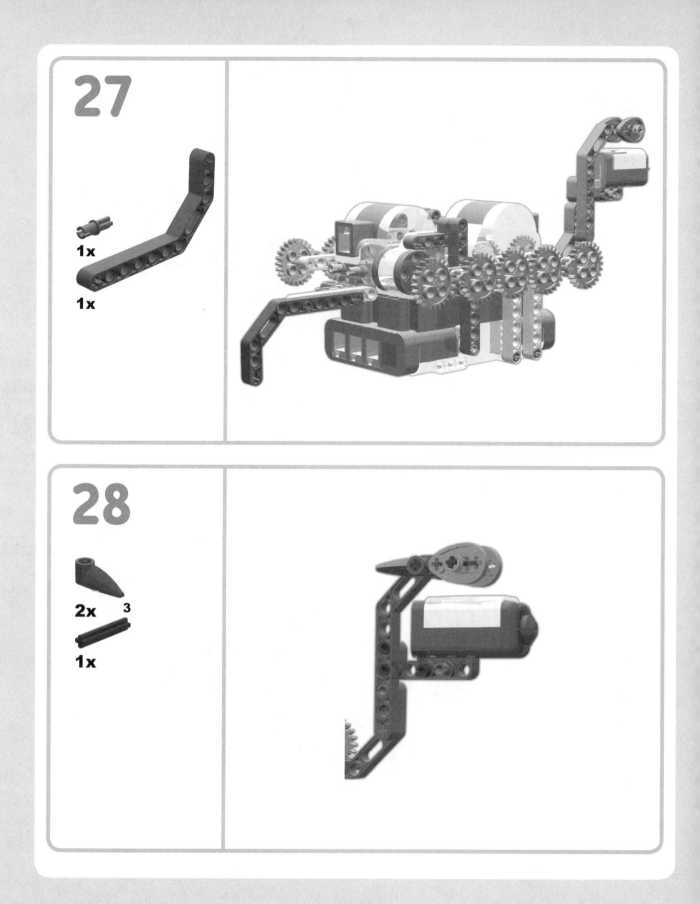

29

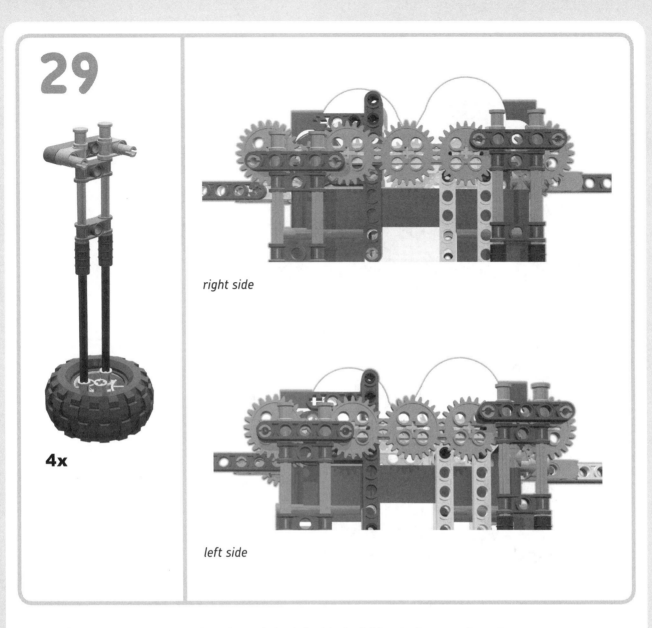

4x

right side

left side

As you can see, the right side and the left side look identical as you face them. Make sure the gears are lined up exactly as shown. (The holes in the gears should line up exactly.)

The leg on the left should be in the bottom-left holes. The leg on the right should be in the top-right holes. The open hole in the wheel should be under the camel.

This is how Sandy should look when he's completed.

wiring connections

Connect the two leg motors to ports B and C. Connect the Touch Sensor to port 2.

programming sandy

We'll create one My Block for Sandy that will simply tell him to walk.

my block: camelwalk

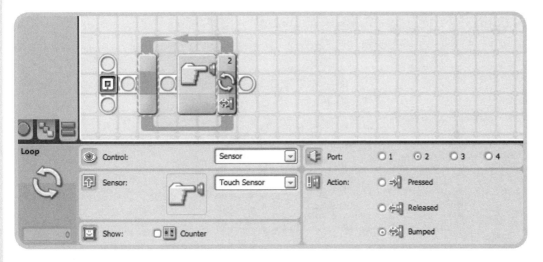

Figure 4-2: Put a Loop block on the beam and configure it to stop repeating when the Touch Sensor is touched. The Touch Sensor is plugged into port 2, so make sure you choose that port when you configure this block.

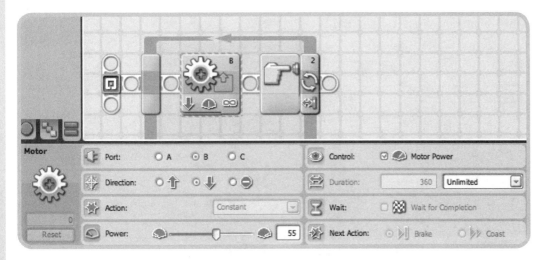

Figure 4-3: Place a Motor block in the loop. Configure it as shown.

To insert the second Motor block on a parallel beam, you have to expand the loop. (This is sometimes referred to as using a crowbar.)

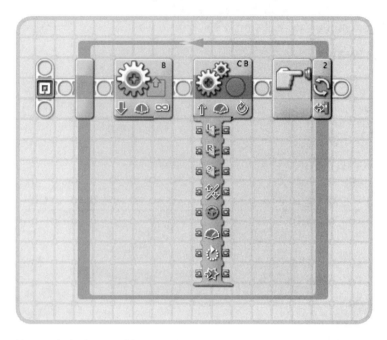

Figure 4-4: Place a Move block in the loop to the right of Motor block B. Expand the data hub by clicking on the lower-left corner of the Move block.

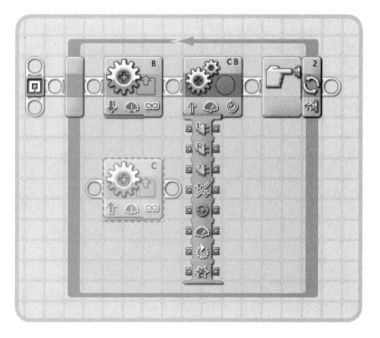

Figure 4-5: Place a second Motor block below the B Motor block.

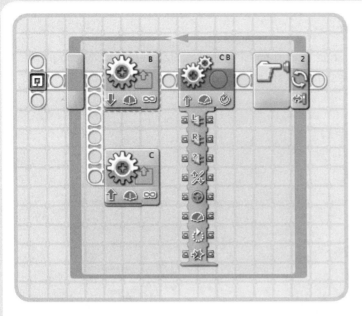

Figure 4-6: Drag a beam from in front of the B Motor block down to the C Motor block.

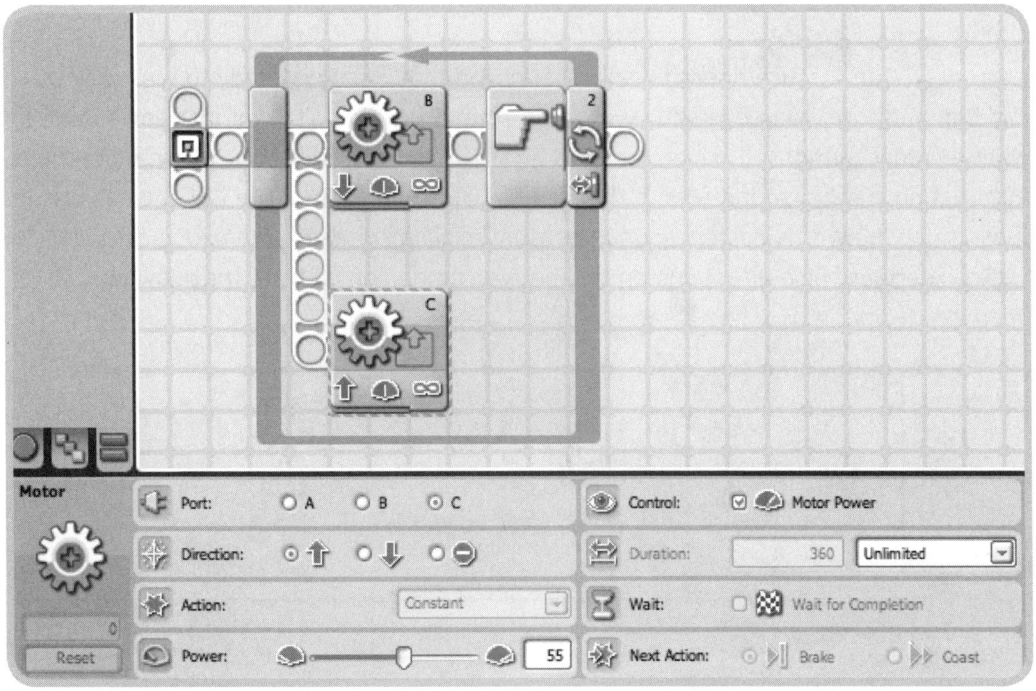

Figure 4-7: Delete the Move block and configure the C Motor block as shown.

Now select the loop and click the Create My Block button on the top Toolbar. (Look for a pair of small blue bars.) Name this My Block *camelwalk*.

the main program

When the brick is facing down under the robot, I make a point to begin my program with a few steps that make it easier to start the robot.

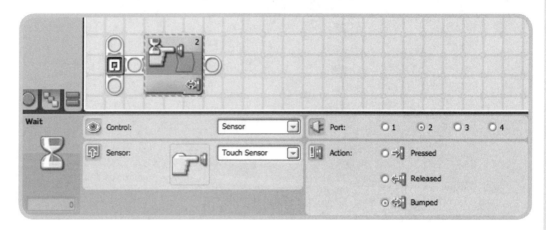

Figure 4-8: The first Wait block is configured to wait for a touch. The Touch Sensor will be connected to port 2, so remember to choose that port in this configuration.

That first touch doesn't really assure me that Sandy is ready to walk unless I add an audio or visual cue—so I will add a sound. But, I don't want just any sound; I want a camel sound, so I have created a brief bit of camel talk. You will find directions for creating custom sound files in the MINDSTORMS section of the LEGO website. (Search for the words *custom sounds* at *http://mindstorms.lego.com/nxtlog/default .aspx*.) There are also directions on this book's companion website (*http://www .thenxtzoo.com/*).

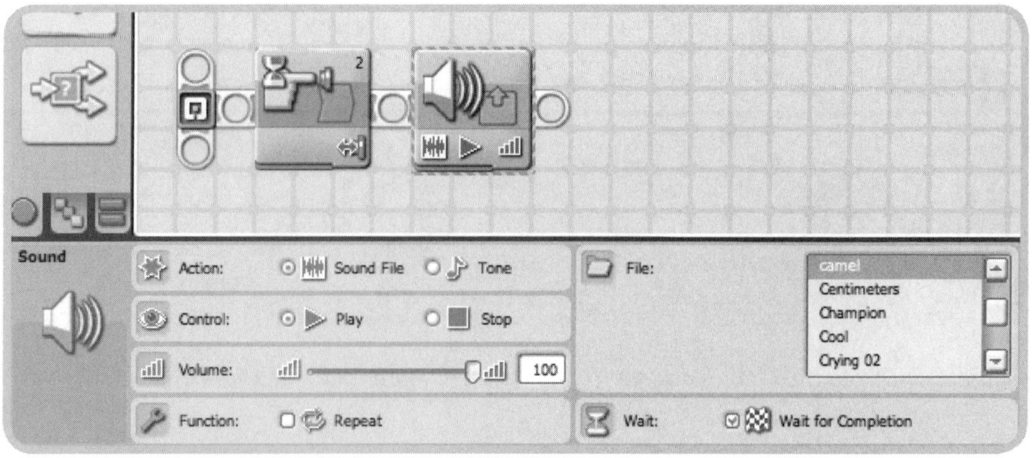

Figure 4-9: Insert a Sound block and choose a sound.

When Sandy tells me he's ready to go by making his special sound, I know he will start walking with one more press of the Touch Sensor. Then he will keep walking until the Sensor is touched again. We make this happen by adding another Wait block configured as a Touch Sensor.

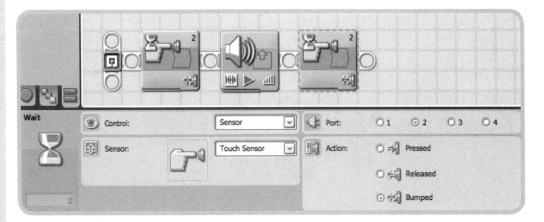

Figure 4-10: This block should be configured exactly the same way as the first Wait block.

Figure 4-11: Insert a camelwalk My Block on the beam.

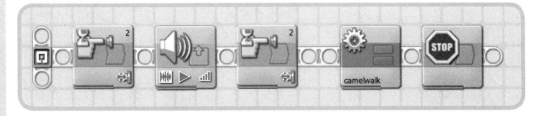

Figure 4-12: Finish the program with a Stop block.

Before you start the robot, recheck the position of the gears and leg attachments. Also, make certain that Sandy's legs are put together tightly and are the same length. Correct alignment is absolutely necessary.

spiderbot: an NXT spider

Spiderbot is a walking eight-legged spider that avoids objects, turns, and runs in forward or reverse.

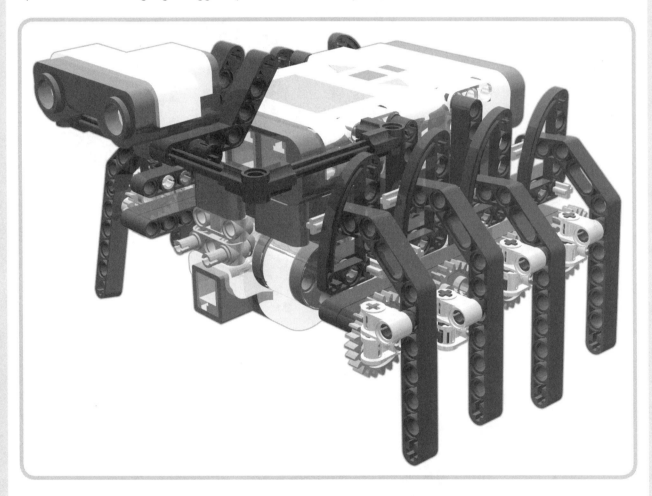

Figure 5-1: Spiderbot!

building spiderbot

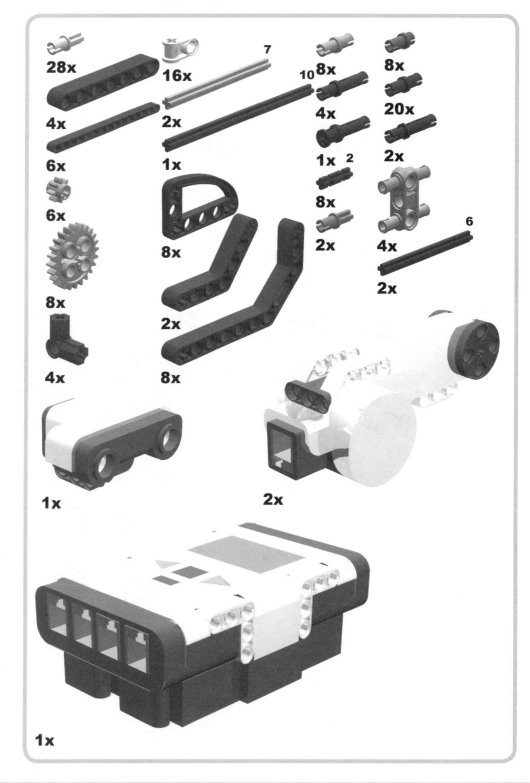

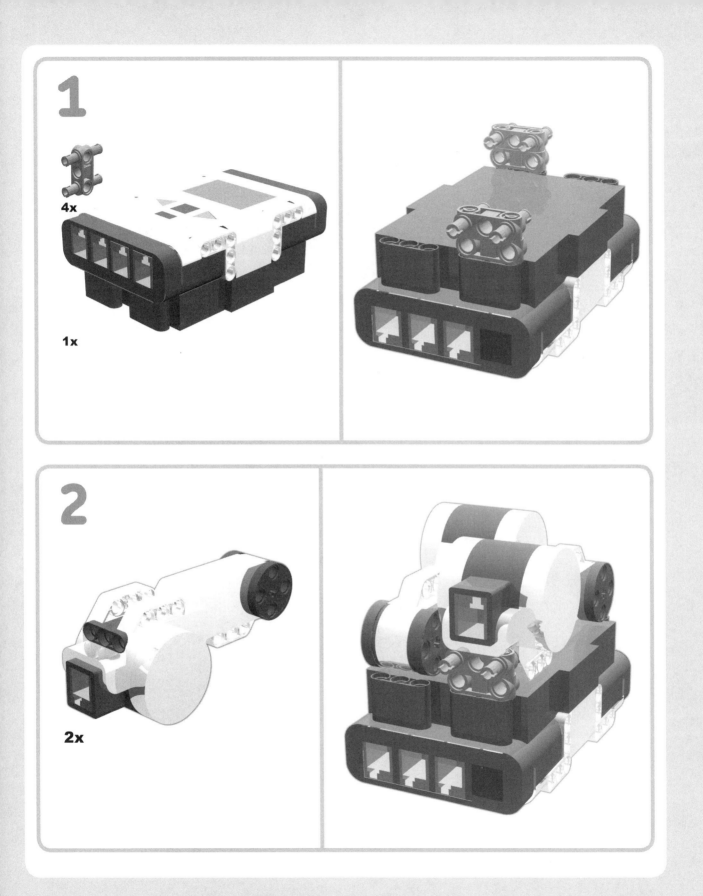

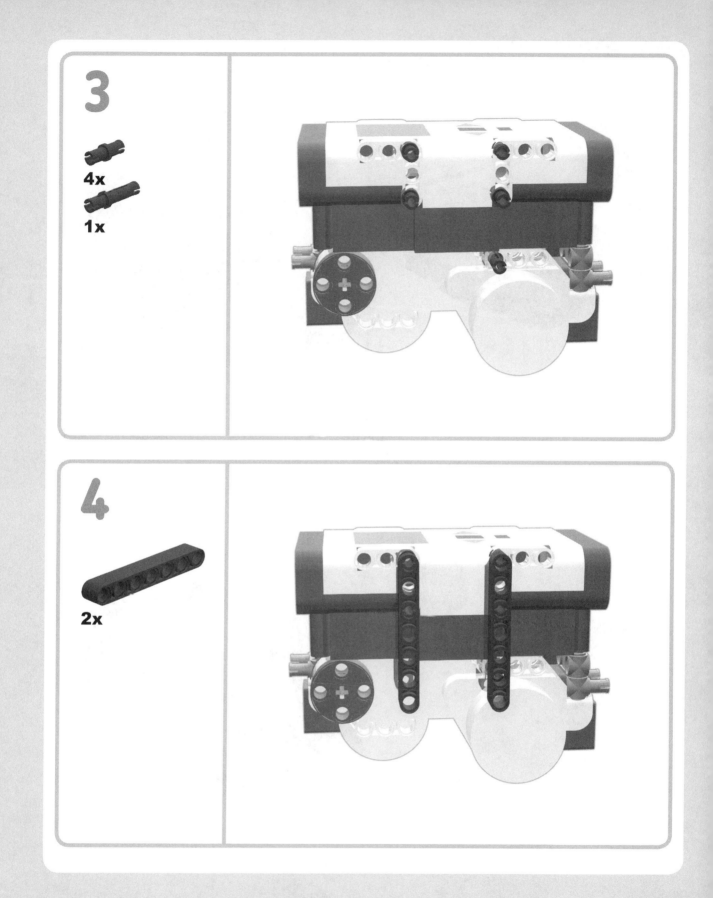

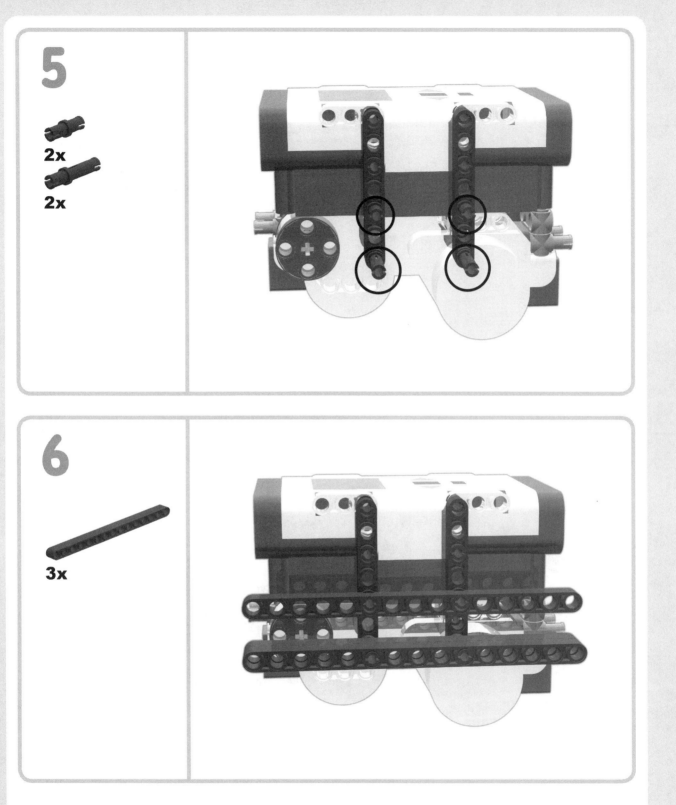

Place two beams on the long pins at the bottom.

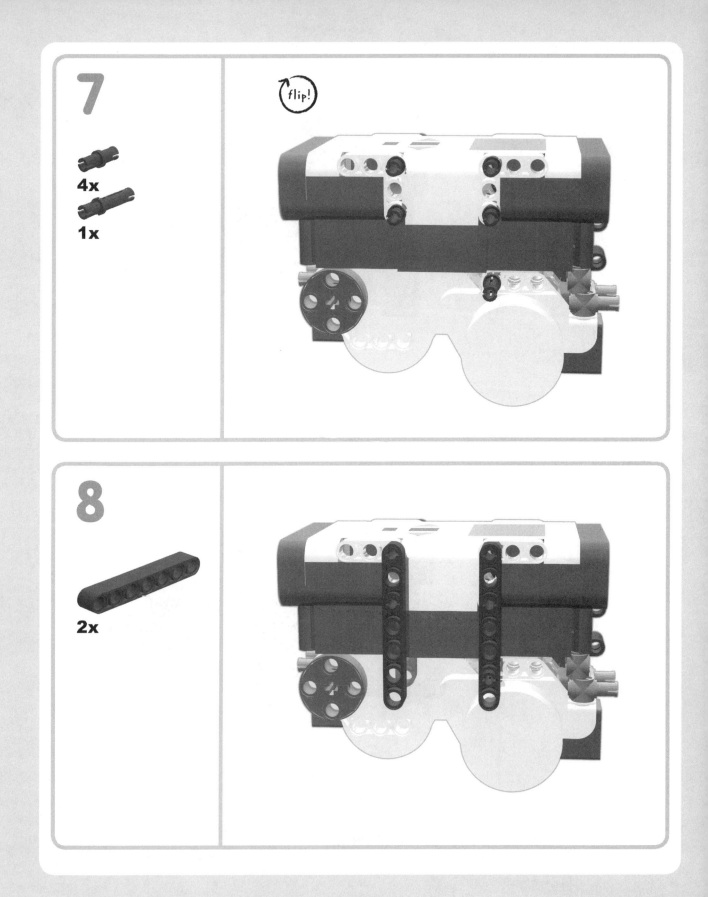

9

2x

2x

10

3x

Place two beams on the long pins at the bottom.

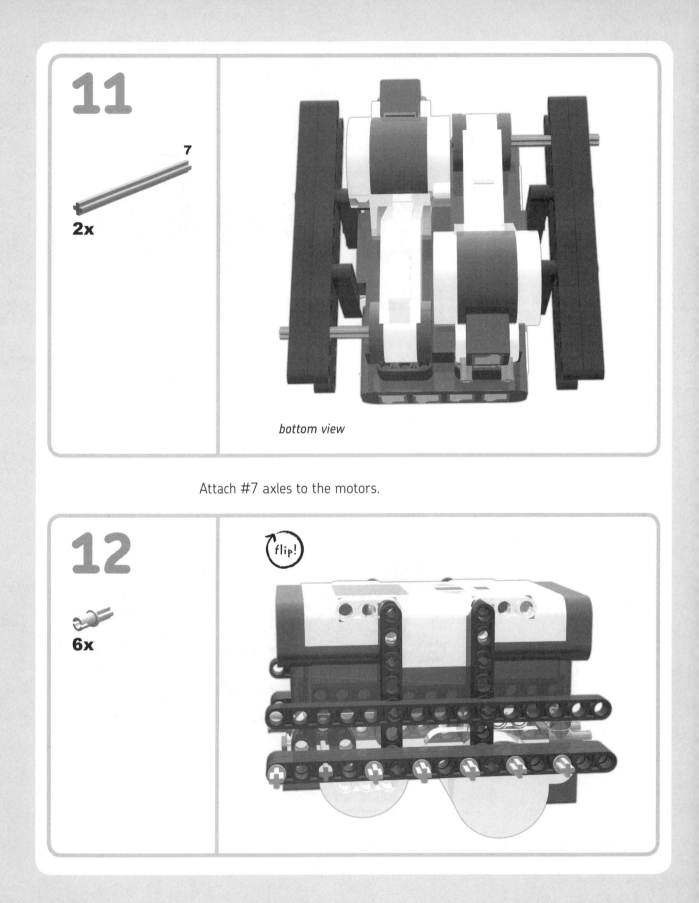

11

7

2x

bottom view

Attach #7 axles to the motors.

12

flip!

6x

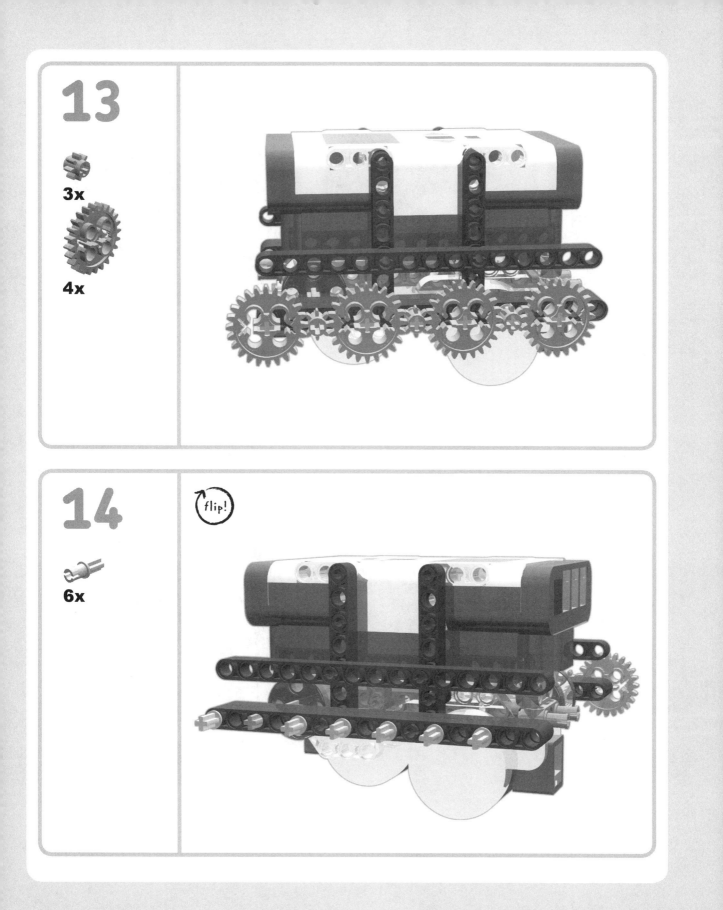

13

3x

4x

14

flip!

6x

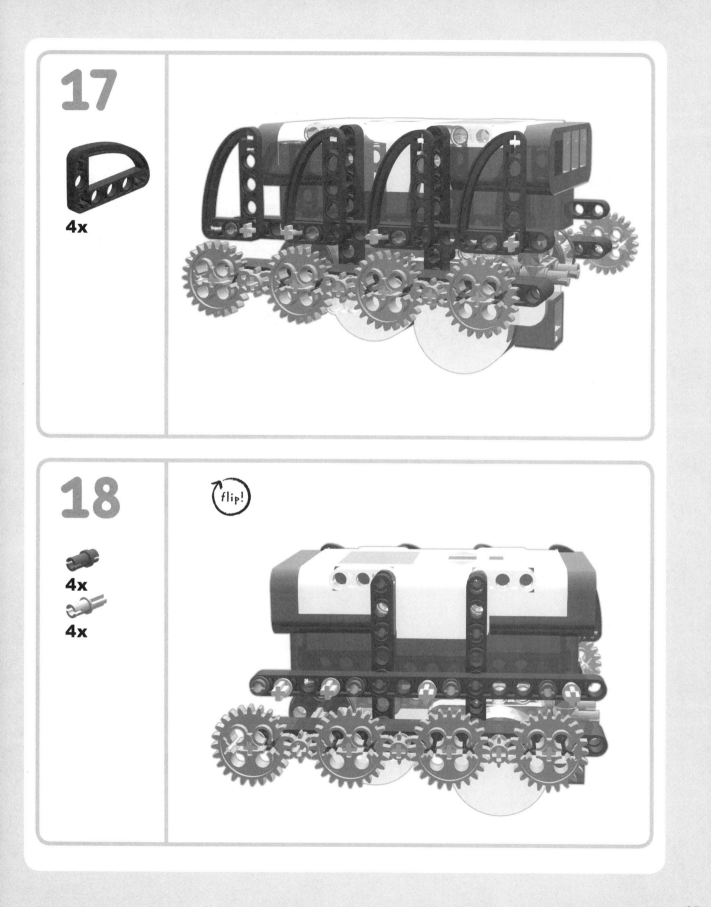

17

4x

18

flip!

4x

4x

19

4x

20

4x 2

4x

4x

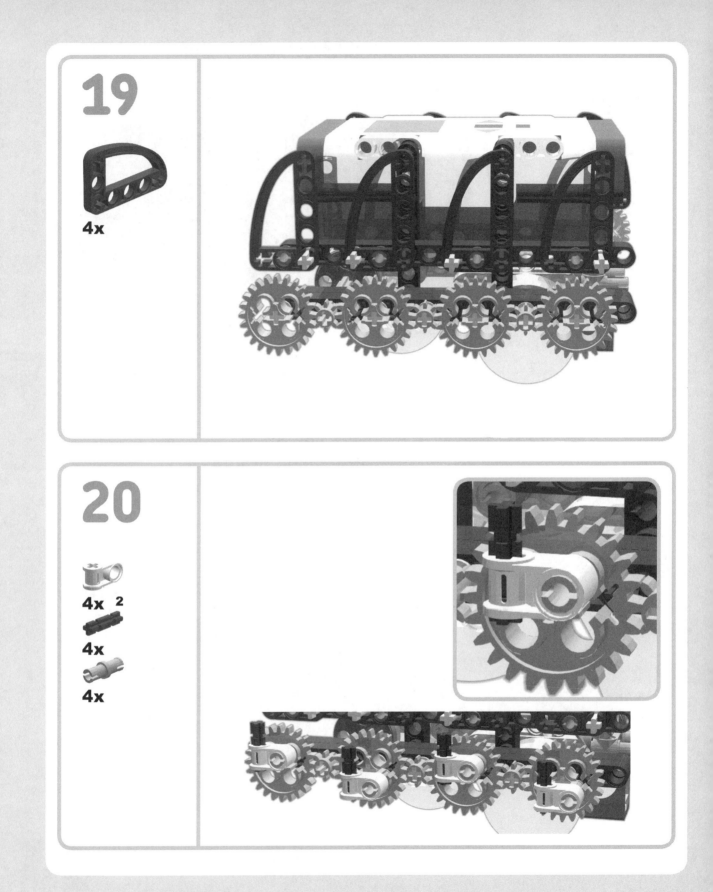

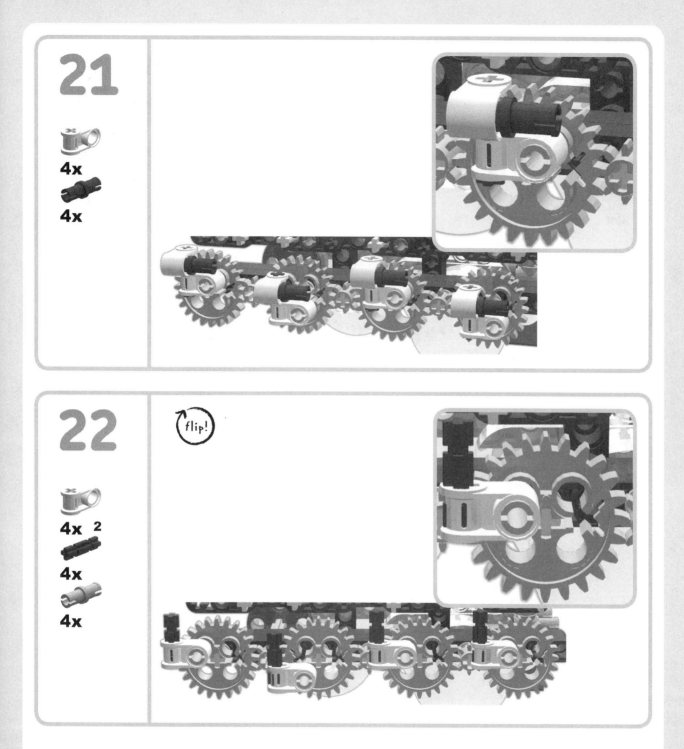

Note that the pins are placed differently on this side than the other side.

NOTE These pictures are only a guide. You might need to rearrange the placement of these pieces on your robot to get smooth motion.

23

4x

4x

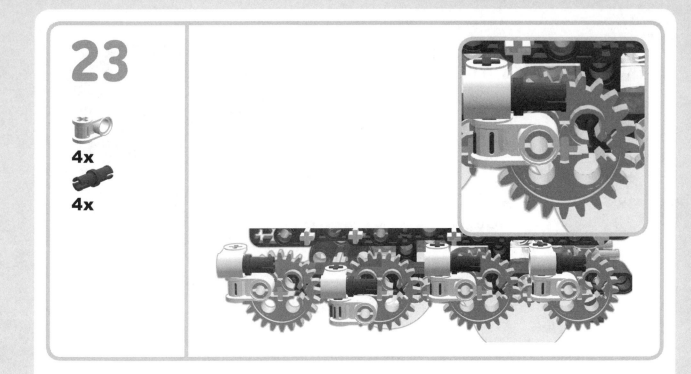

24

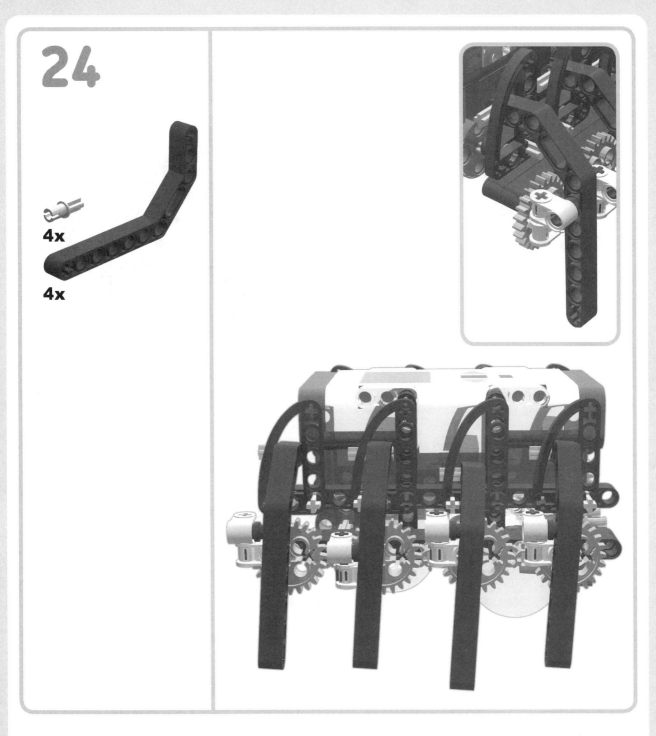

4x

4x

In step 24, insert an axle pin (any color) at the end of the beam. Turn it at a right angle to slide it through the supporting "shoulder" beam. The pin will keep the beam from coming out of the shoulder piece.

25

flip!

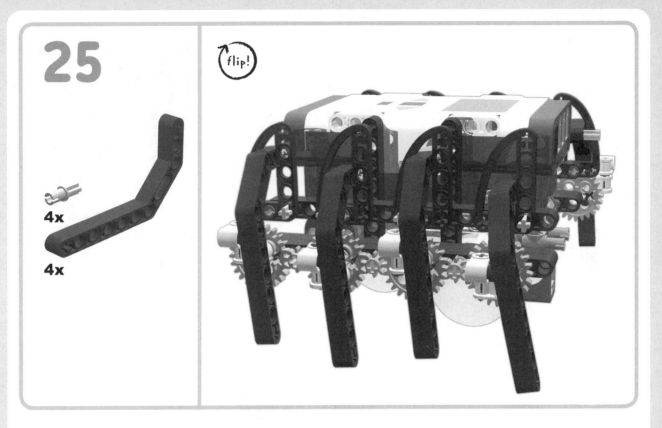

Insert an axle pin (any color) at the end of the beam. Turn it at a right angle to slide it through the supporting "shoulder" beam. The pin will keep the beam from coming out of the shoulder piece.

pause now to test your robot's legs

Now is the best time to test Spiderbot's legs. Turn the spider on its back and remove the driving gears (the small gears on the motor axles). Now you can spin all the gears both ways to discover problem areas and make adjustments. When all the legs and gears move freely, reinsert the driving gears and move on to the next step.

NOTE Keep in mind that you might need to change the placement of the legs to find the best arrangement for your robot.

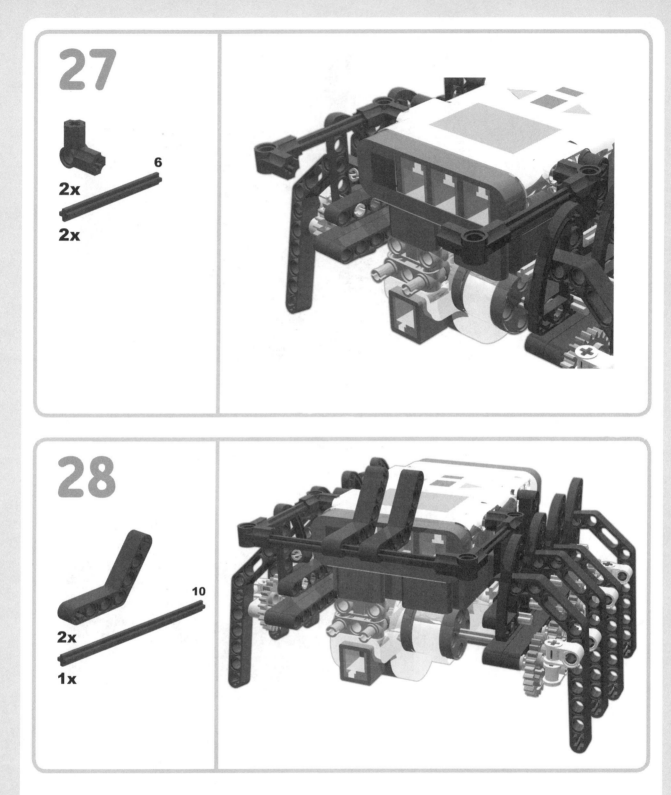

Step 28 shows you where these parts will go. Wait to assemble them with the Ultrasonic Sensor in the next step.

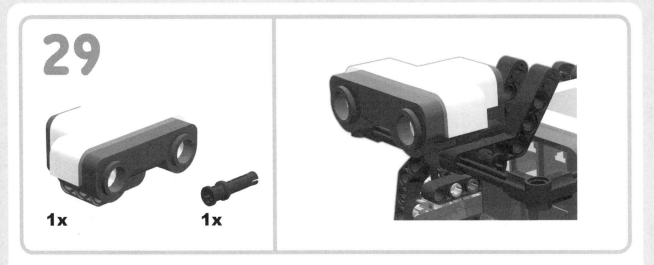

29

1x   1x

Place the sensor between the two bent beams. Thread the axle through the first hole in the sensor; insert a long pin into the second or third hole.

wiring connections

Connect the motors that drive the legs to output ports B and C. Connect the Ultrasonic Sensor to input port 4.

programming spiderbot

Let's write a program to make the spider move around the room while avoiding objects in front of it.

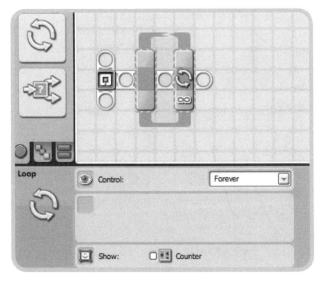

Figure 5-2: First, we'll place a Loop block on the beam and configure it as a Forever loop. This loop will contain all the action steps for the spider.

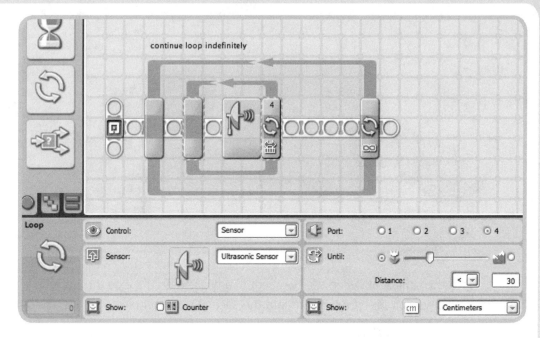

Figure 5-3: Place another Loop block inside that loop, but this time, configure it as shown here: an Ultrasonic Sensor looking for a distance of less than (<) 30 centimeters. Input port 4 is the usual choice for the Ultrasonic Sensor.

Because the robot has motors placed in opposite directions, we need to choose different directions for each of the motors in the program.

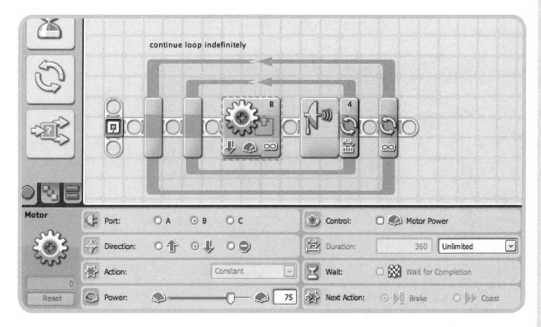

Figure 5-4: Place a Motor block in the Ultrasonic Sensor loop and connect it to output port B. Copy the settings displayed in the configuration panel shown here.

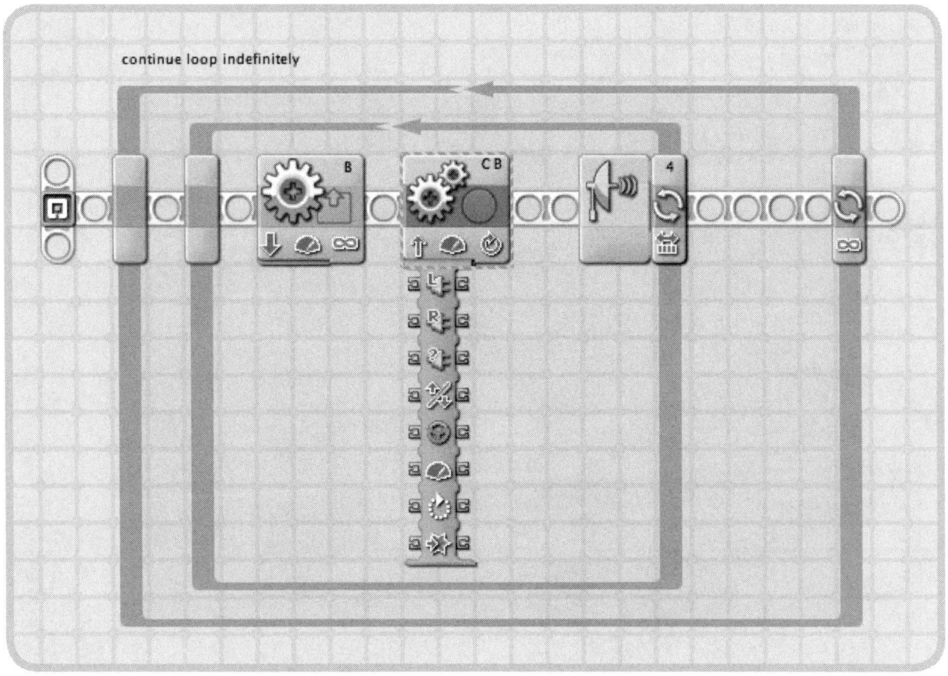

Figure 5-5: Here is another place where we need to use the expansion trick from Chapter 4 (pages 80–81). Add a Move block to the loop and click the lower left edge of the block to open the data hub. (The data hub is the bar hanging down from the Move block.)

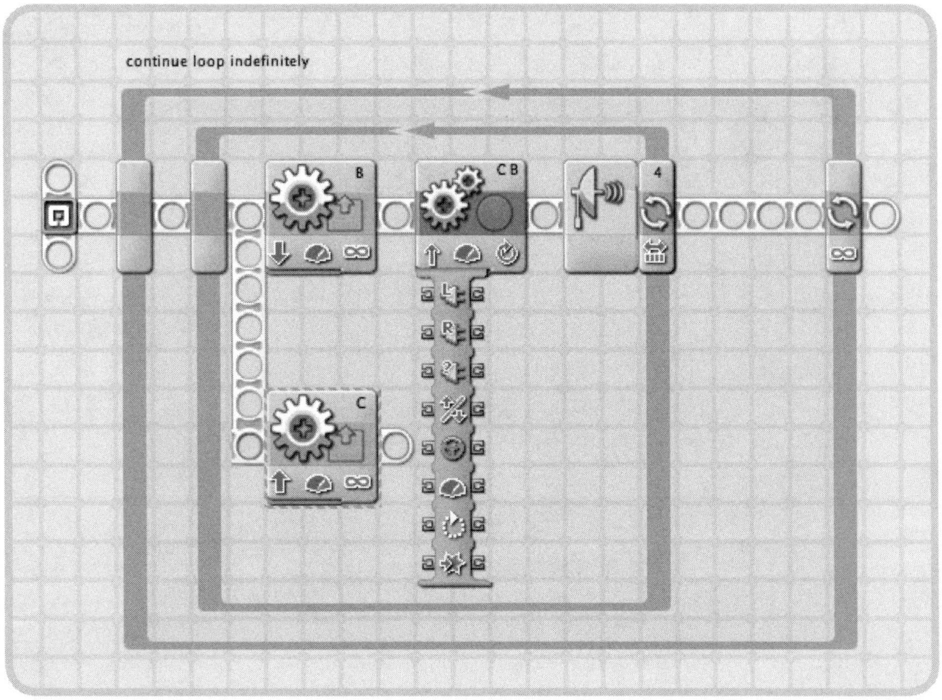

Figure 5-6: Insert a Move block below the B Motor block. Press SHIFT and use your mouse to drag a beam from in front of the B Motor block down to the C Motor block.

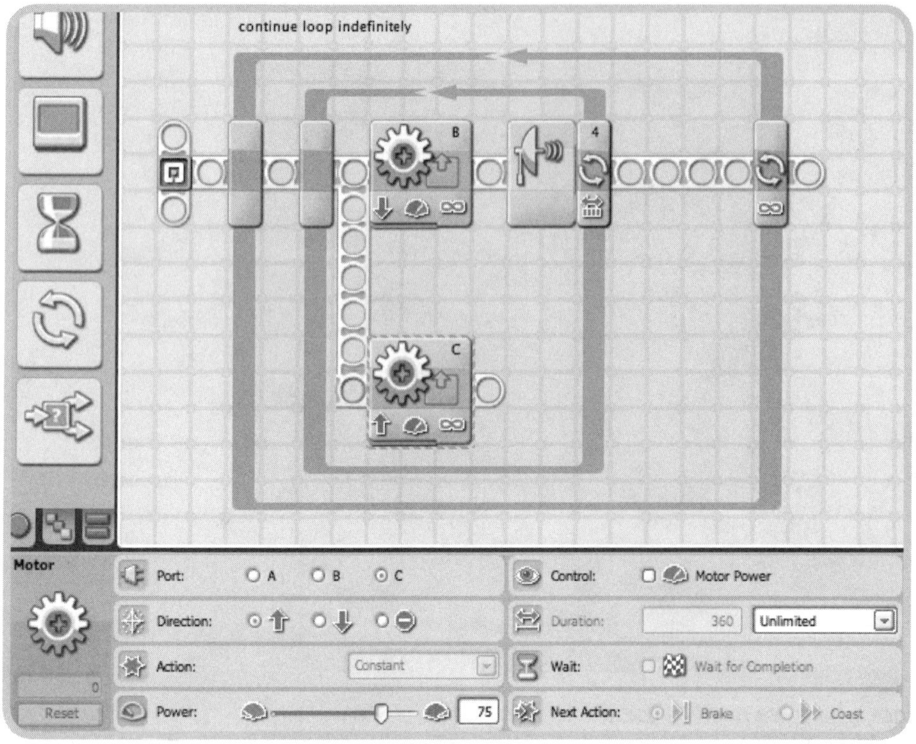

Figure 5-7: Delete the Move block. Settings for the C Motor block are shown here.

To make the robot back up, we'll add two Motor blocks.

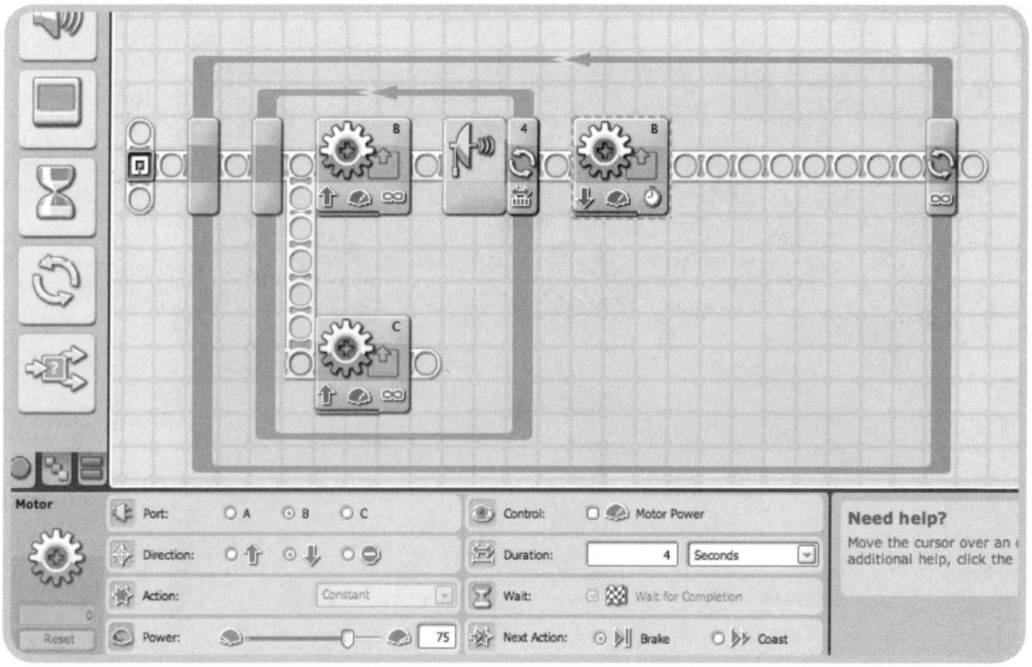

Figure 5-8: First add a Motor block after the inner loop. Select the settings shown here.

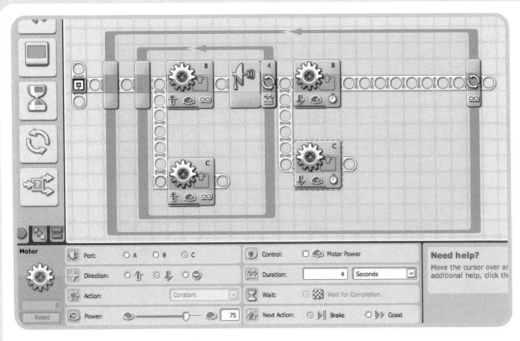

Now we'll place the final Motor blocks. This time we want the spider to turn, so we'll direct both motors to turn in the same direction.

Figure 5-9: Add a Motor block under the B Motor block. Use the settings shown here.

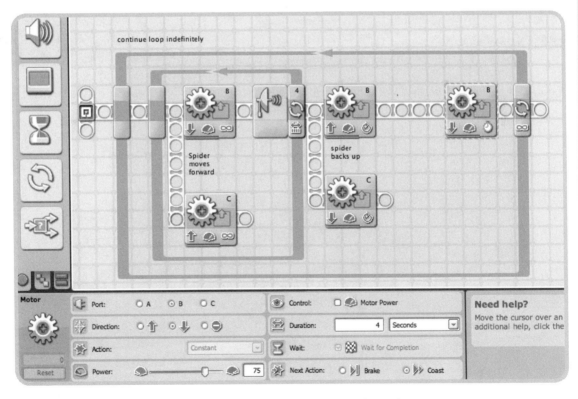

Figure 5-10: Configure the B Motor to move for 4 seconds, as shown here.

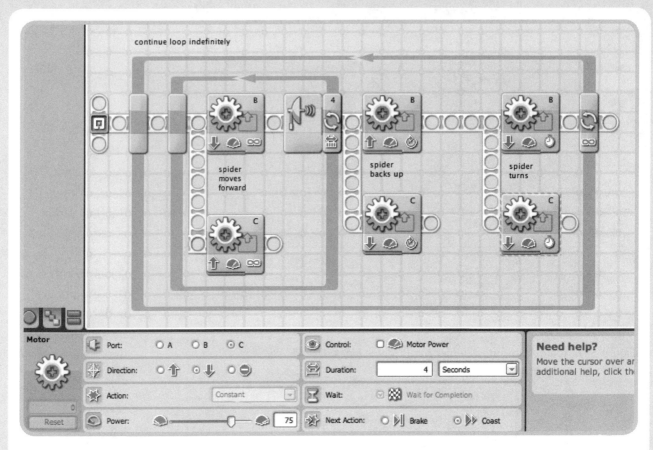

Figure 5-11: Place the second Motor block below the B Motor block and configure it as shown here.

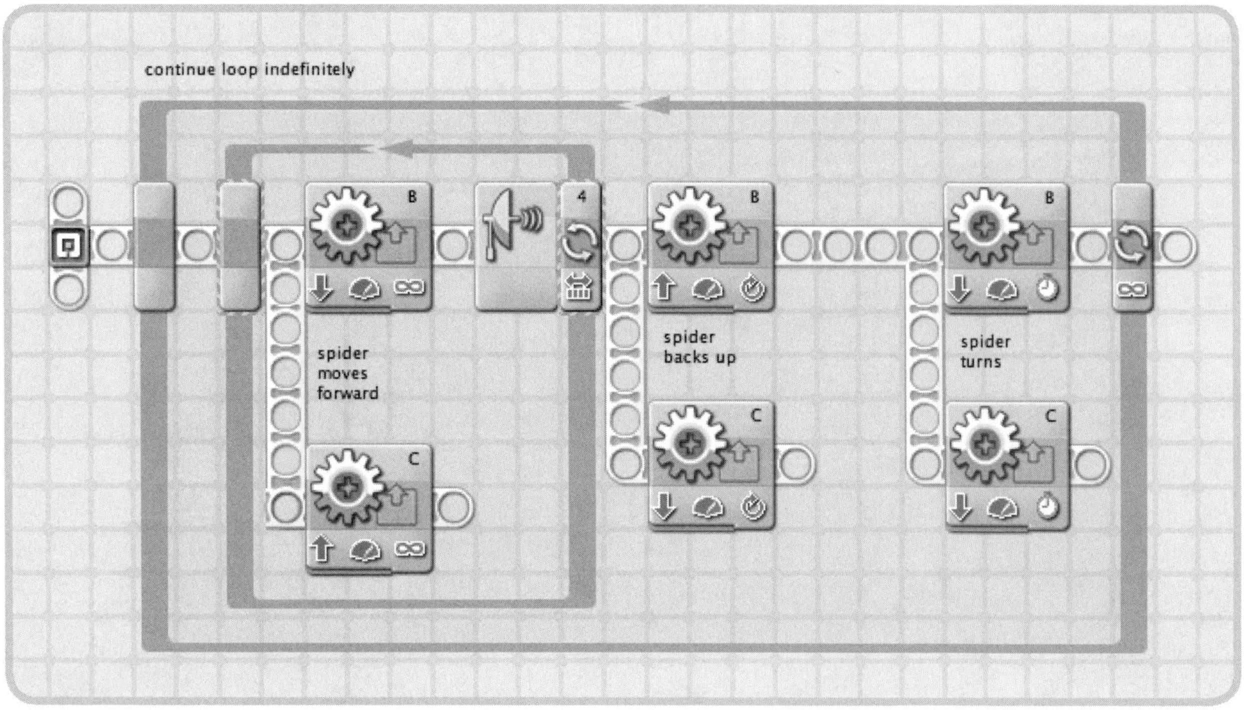

Figure 5-12: The completed program

If you stopped to test and readjust the spider after step 25, you should be ready to download the programs and get your spider crawling.

NOTE If your spider moves in the wrong direction when you start it, either switch the direction of each motor in the program or switch the wire connections to the opposite motors.

snout: an NXT 'gator

One of the things I like best about this robot is that when you set it into motion, it actually looks like an alligator walking. Its jaws also open and close, which adds to the effect.

This is the only robot in which I have used 40-tooth gears as part of the walking mechanism. I find that when using these large gears, the legs must be short and the feet must be broad. Something to remember when you are building your own walking robots is that the longer the legs and the larger the gears, the more problems you will have. All legs put pressure on the gears, but large gears flex more, causing friction with other parts—and that is always bad for moving assemblies.

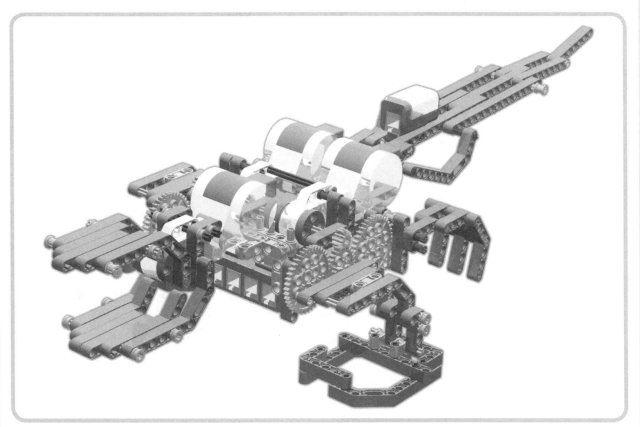

Figure 6-1: The NXT 'Gator

Note that the back legs are identical and can be placed on either side of the robot. The left and right front legs, however, are designed specifically for their locations; they are not the same.

building snout

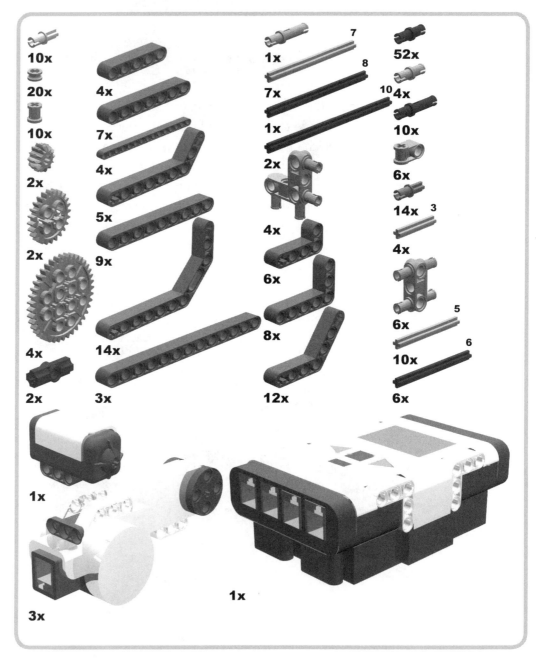

base

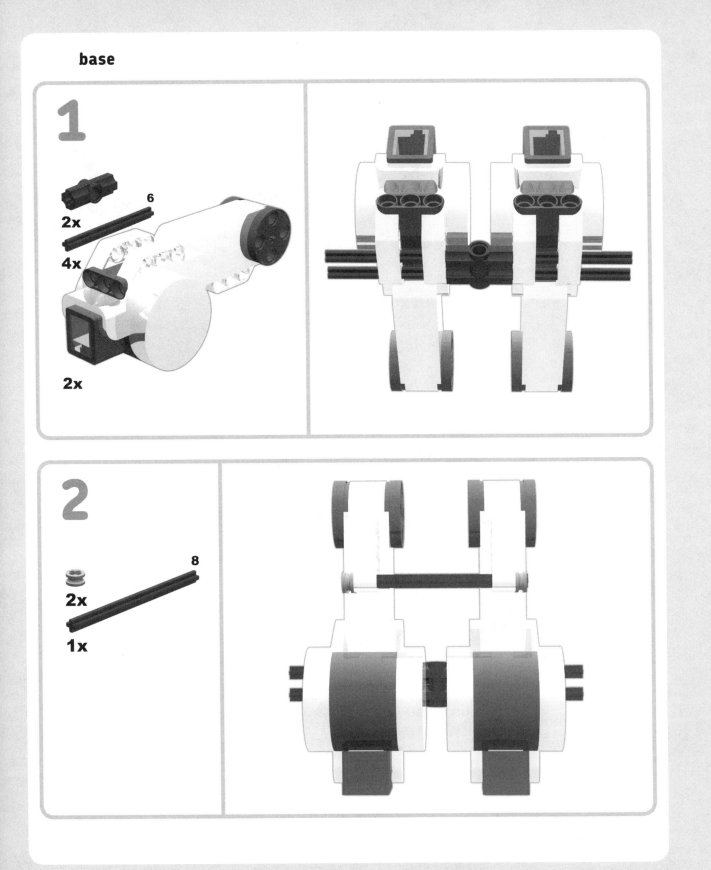

3

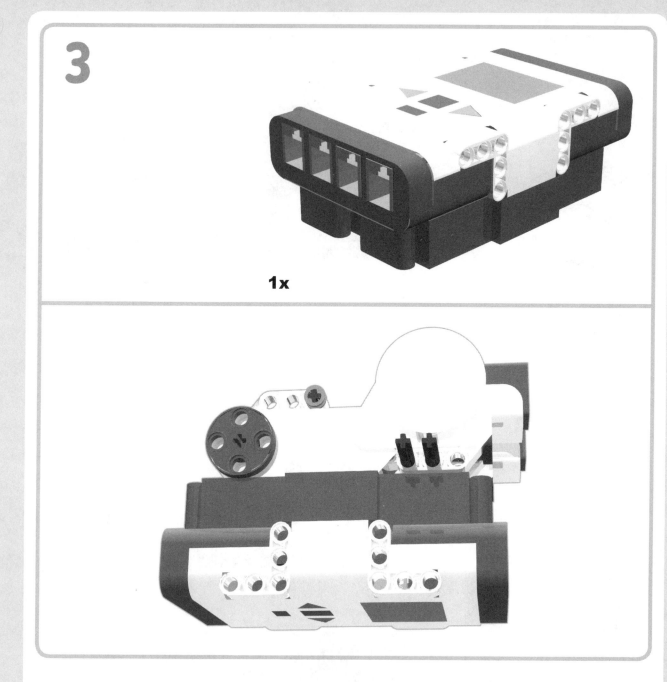

1x

Place the motors on the bottom of the brick, as shown.

4

2x
2x
1x

2x
4x
1x

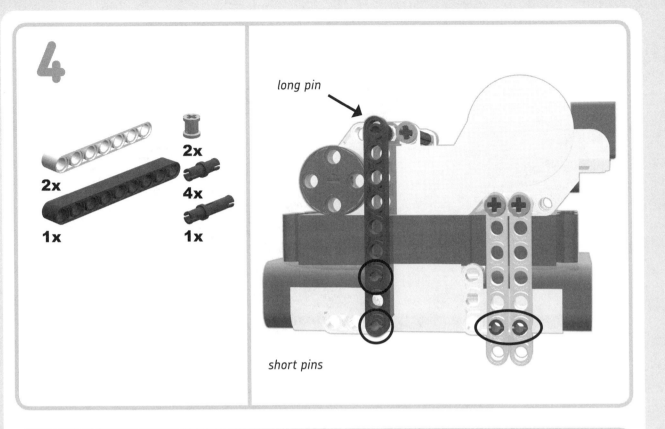

long pin

short pins

5

2x
2x
1x

2x
4x
1x

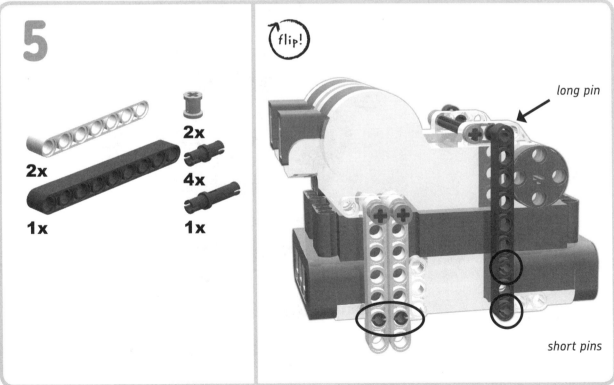

flip!

long pin

short pins

6

2x

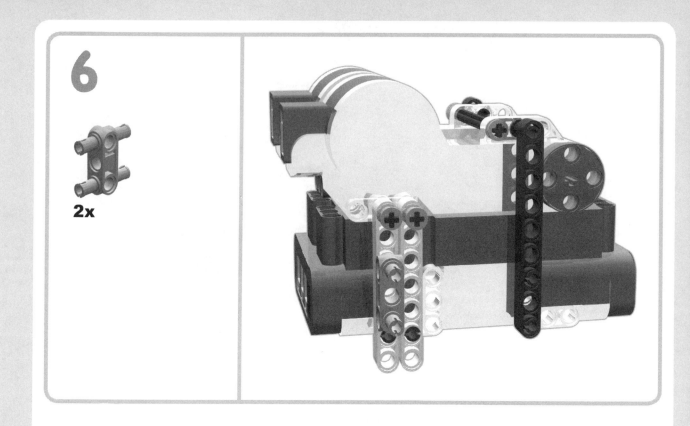

Attach one joiner on each side.

7

7

2x

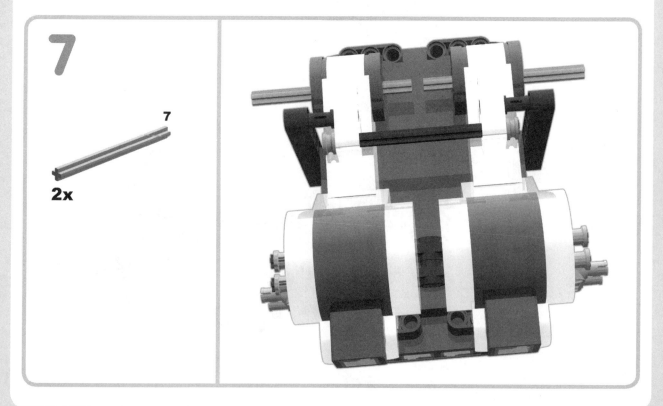

8

4x

2x

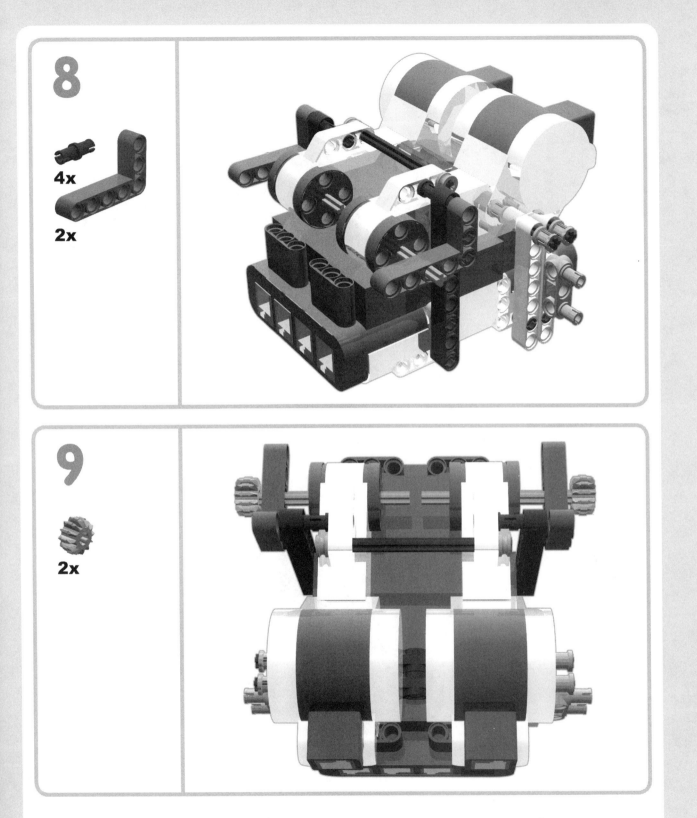

9

2x

These will be the *driving gears* (the gears that make all the other gears move).

10

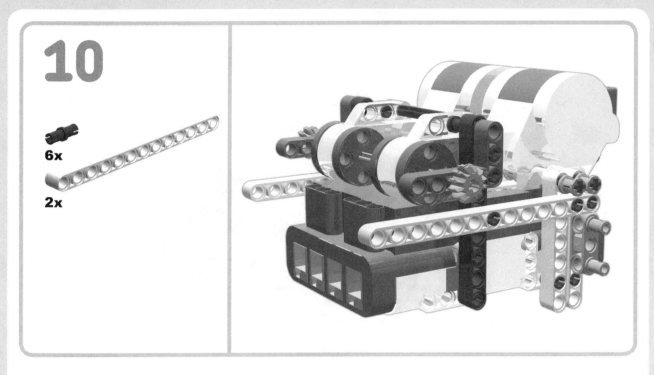

6x

2x

Attach these pieces the same way on each side.

11

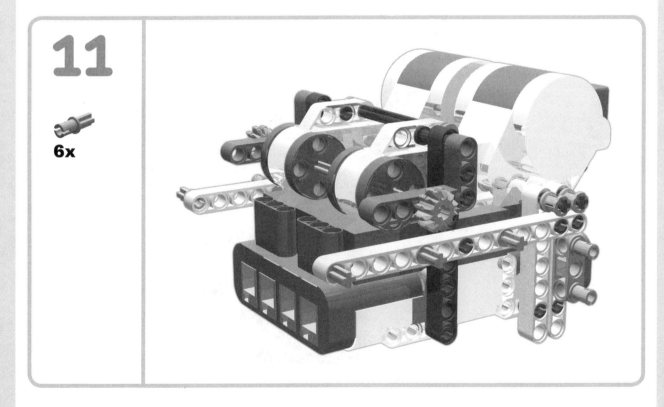

6x

Attach three of these yellow axle pins on each side.

12

2x

4x

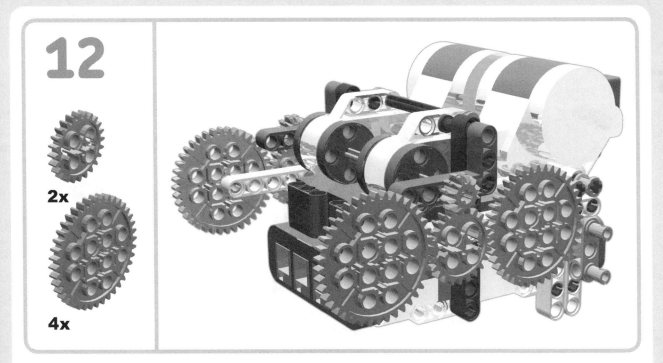

Attach the gears the same way on each side.

13

2x

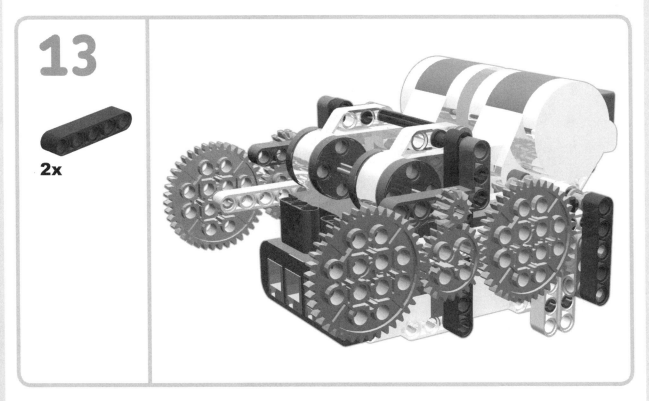

Attach one beam on each side.

14

4x

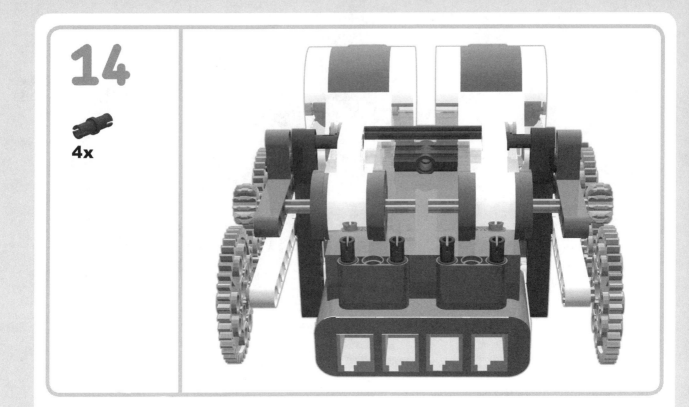

back feet (build two)

1

3

1x

2x

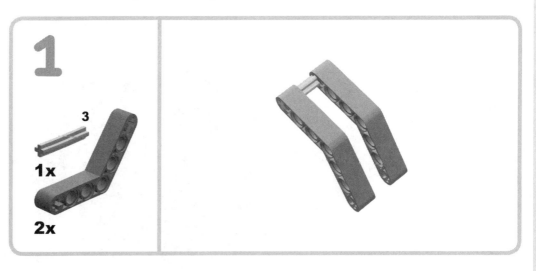

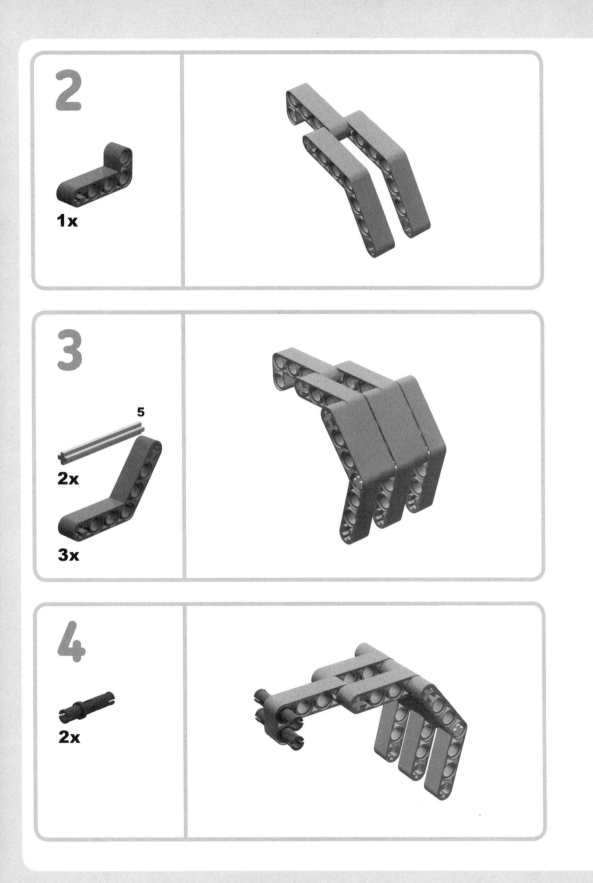

head

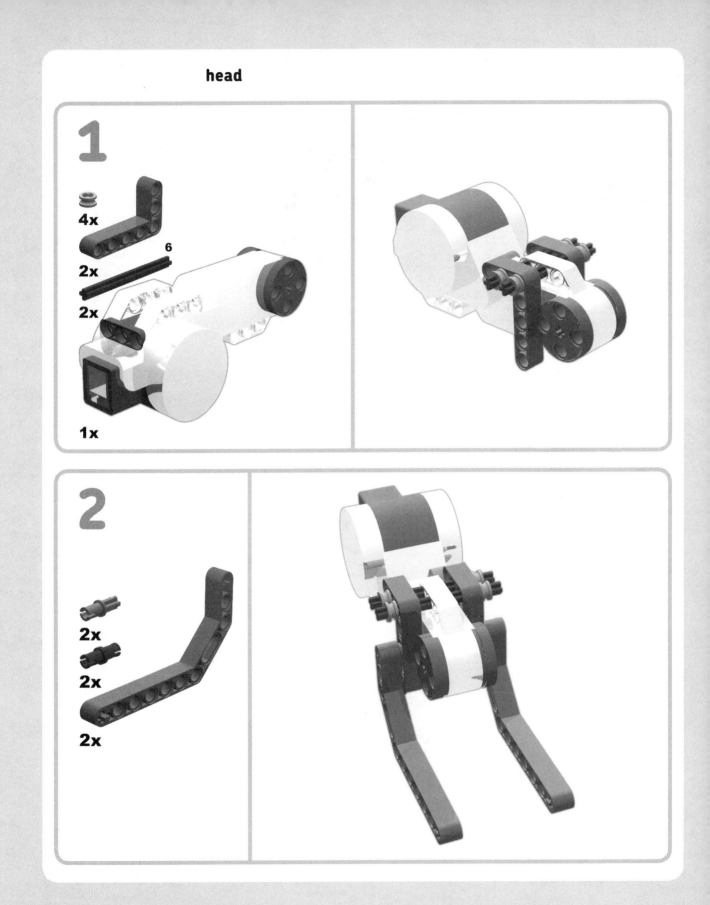

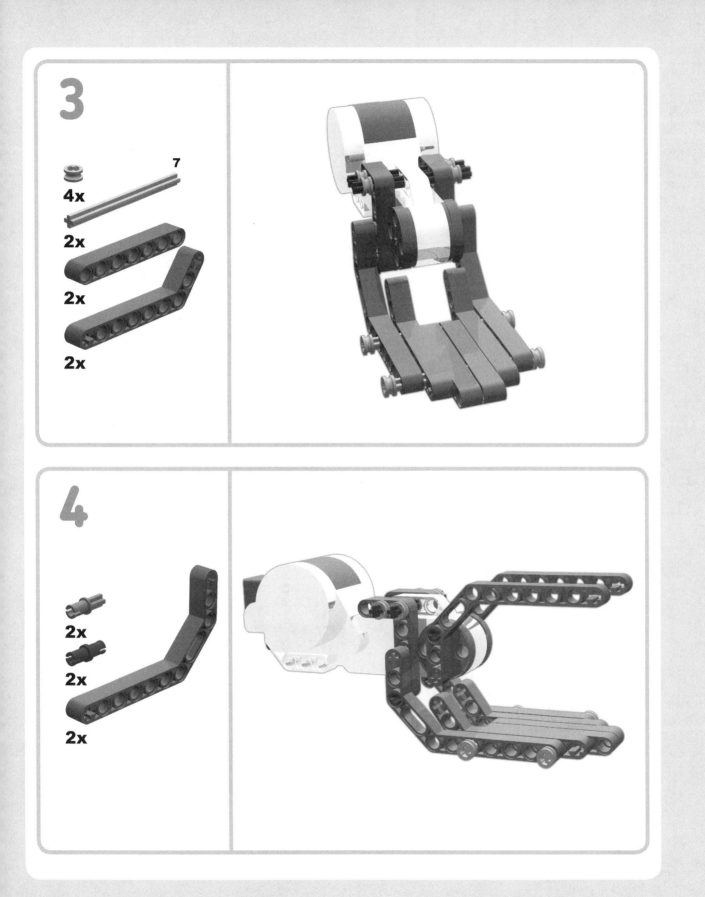

5

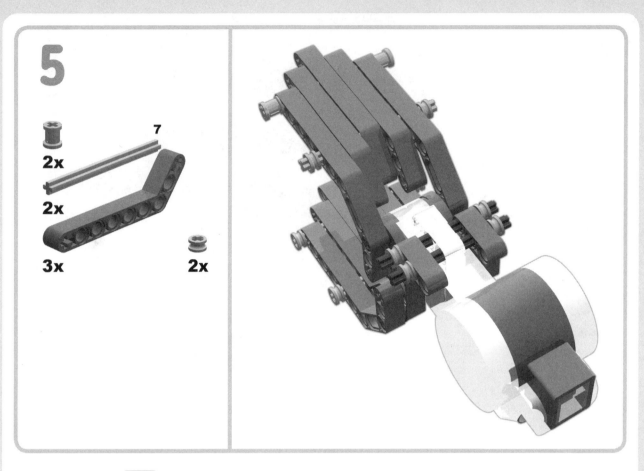

2x

2x 7

3x 2x

NOTE If you have yellow bushings (from the Education Resource Set), use them for the eyes and put two on each side. It looks very cool.

6

2x

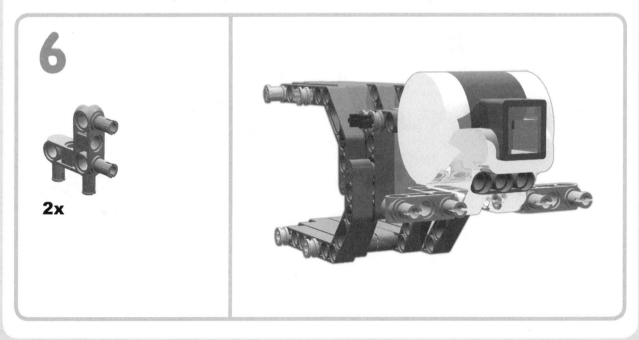

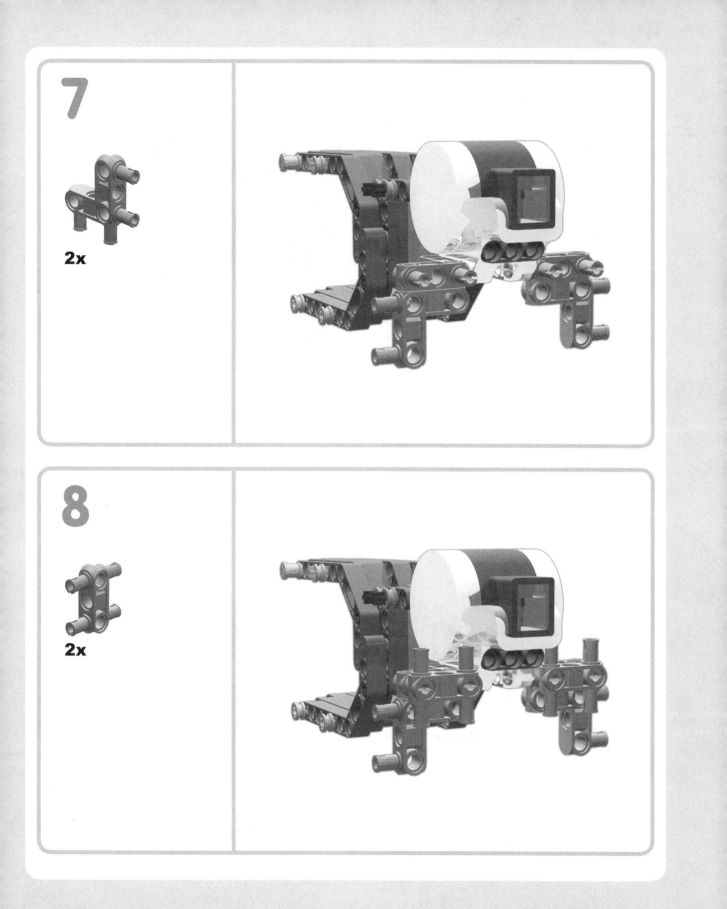

7

2x

8

2x

9

2x

1x

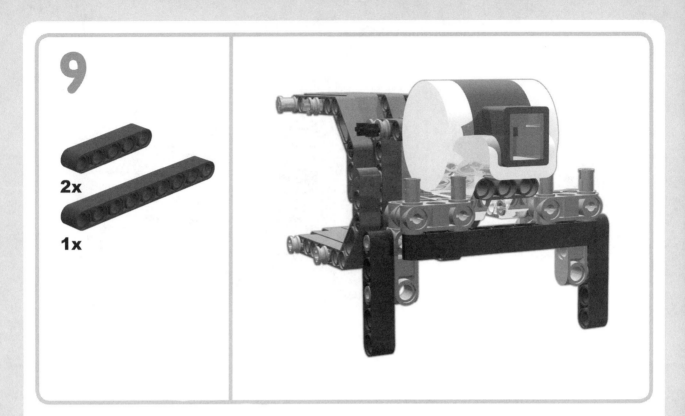

left leg

1

4x

4x

2x

2x

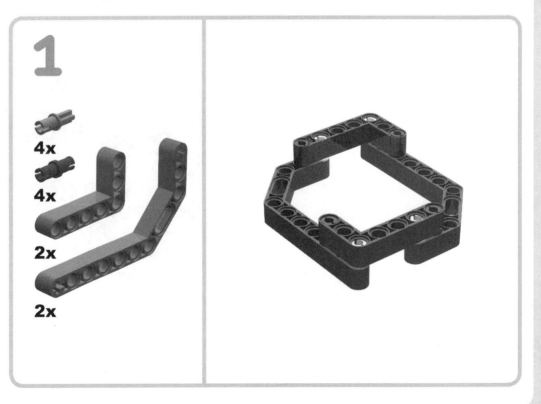

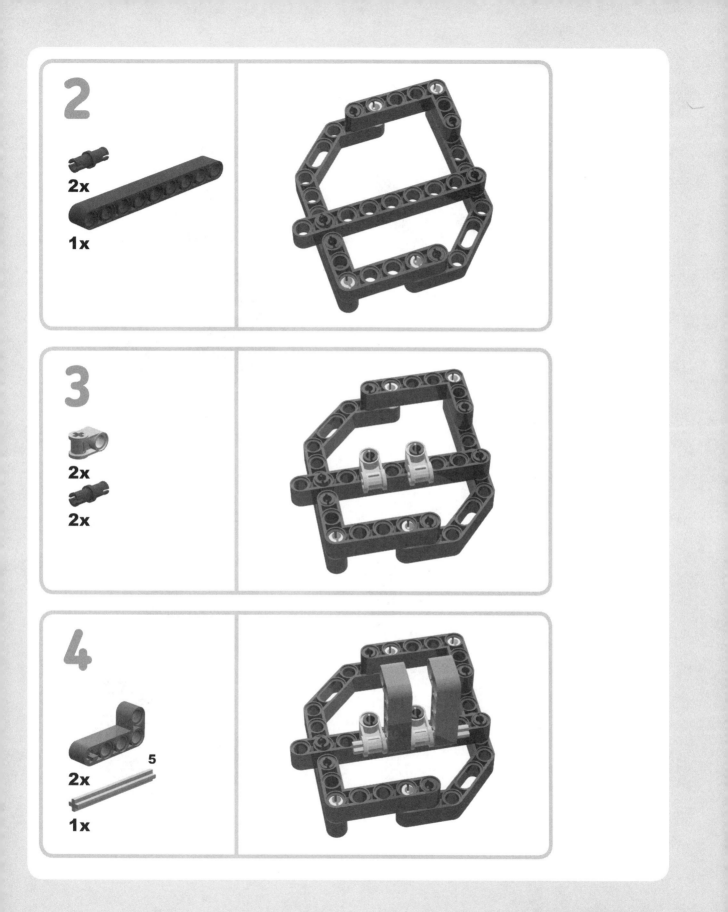

7

4x

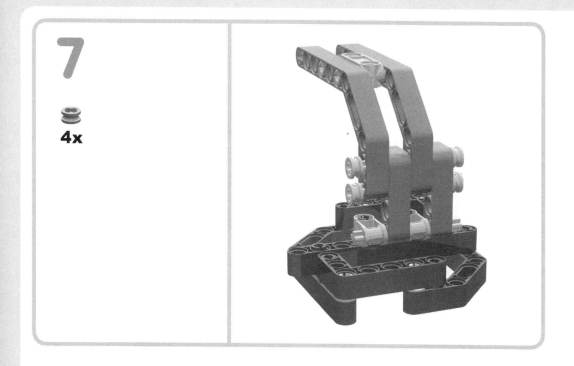

PART SUBSTITUTION SUGGESTION

If you are running short on the thinner half-bushings (Figure 6-2), try using regular bushings (Figure 6-3). Note that this may require using a slightly longer axle.

Figure 6-2: Half-bushing

Figure 6-3: Regular bushing

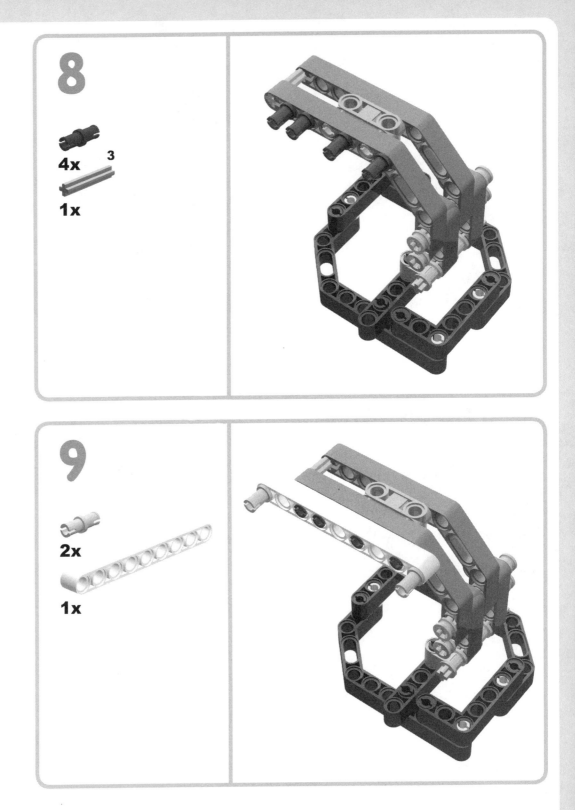

8

4x

3

1x

9

2x

1x

Later on, these gray pins will be inserted into the walking gears.

right leg

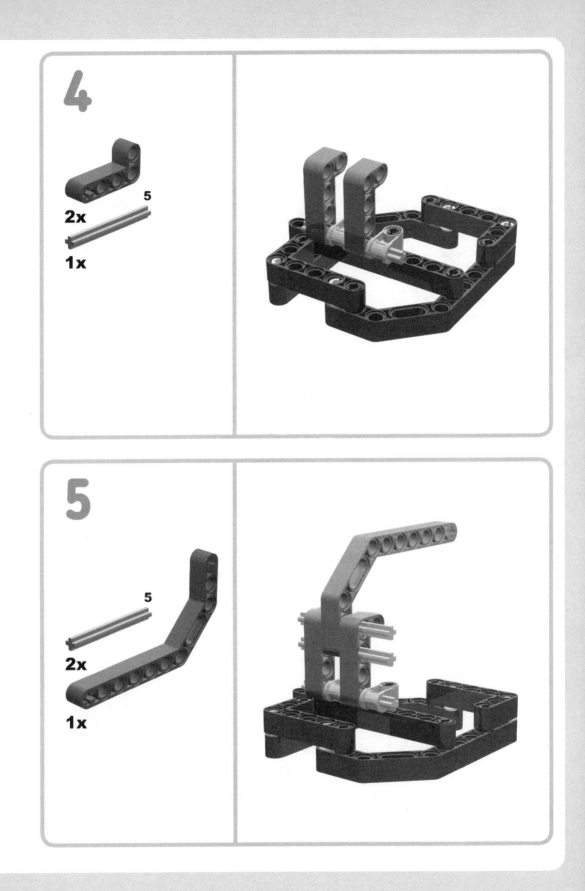

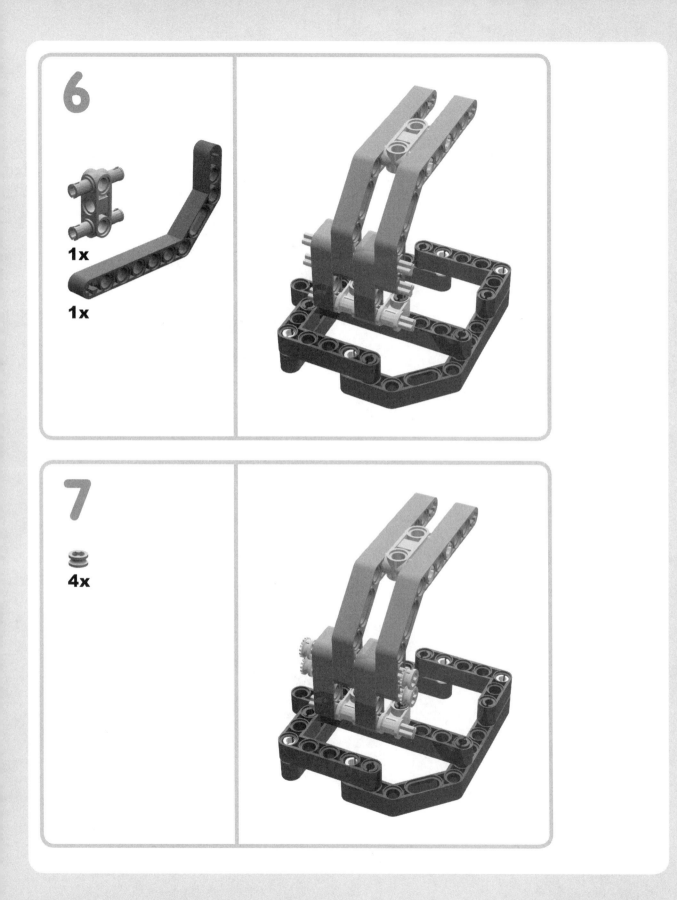

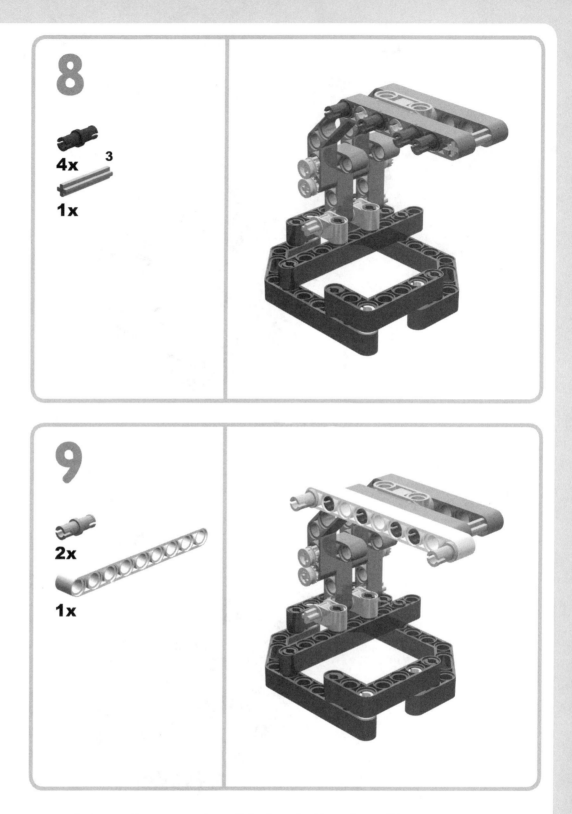

8

4x 3

1x

9

2x

1x

Later on, these gray pins will be inserted into the walking gears.

tail

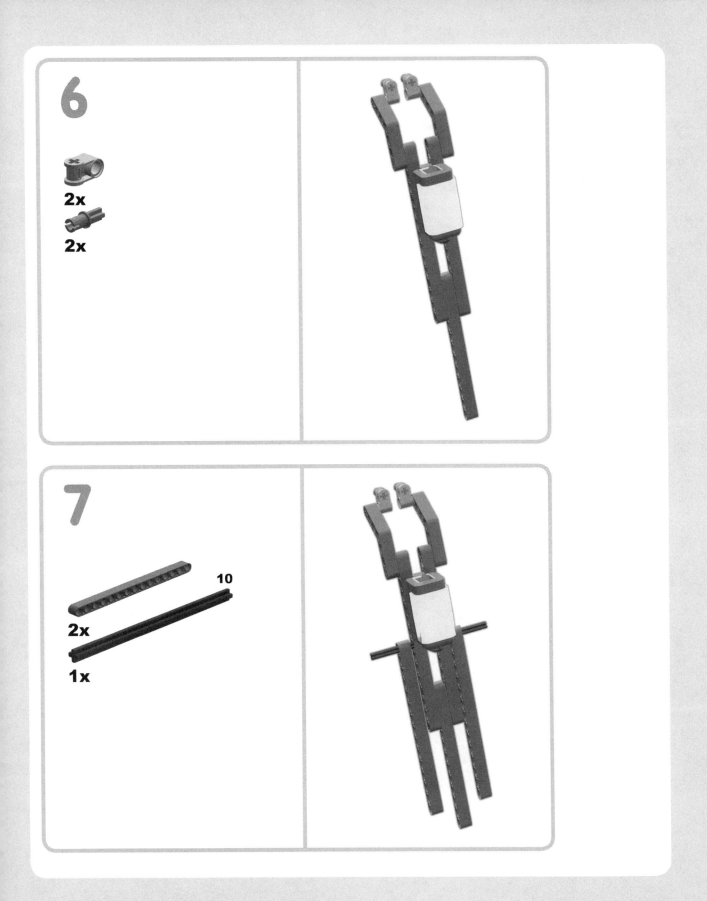

10

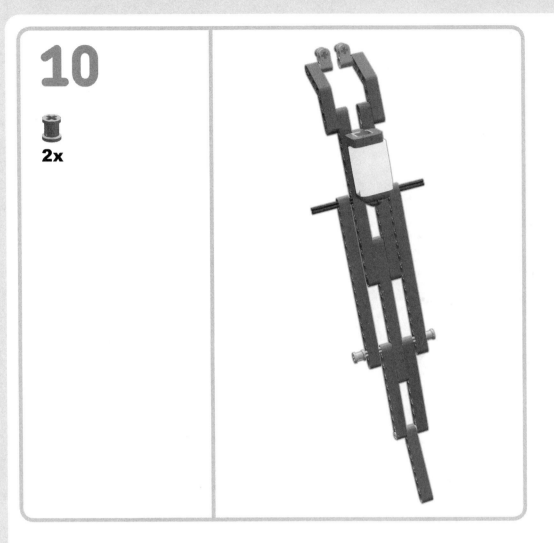

2x

We'll finish the tail after we assemble the rest of the robot.

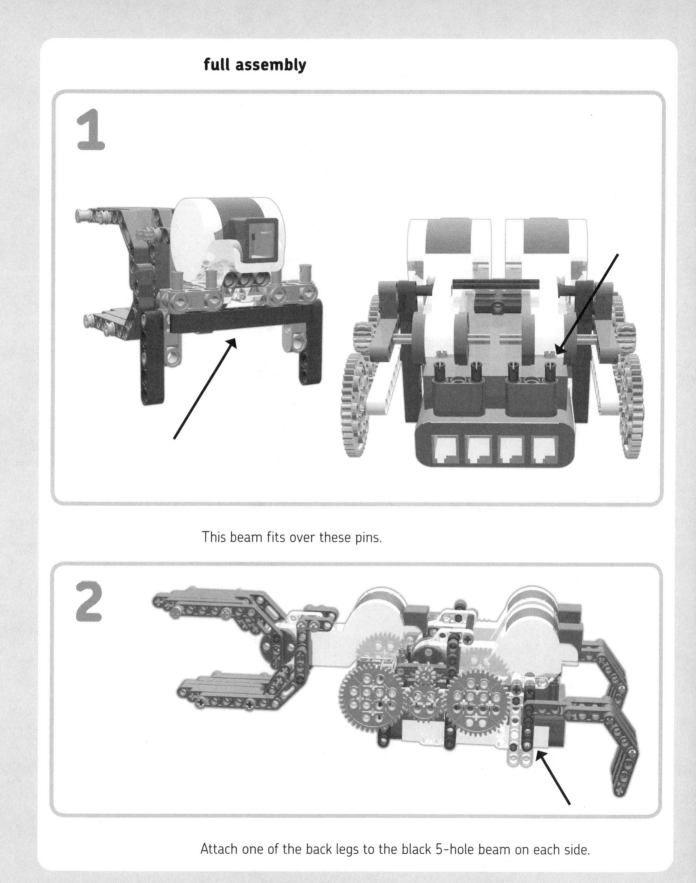

1

This beam fits over these pins.

2

Attach one of the back legs to the black 5-hole beam on each side.

3

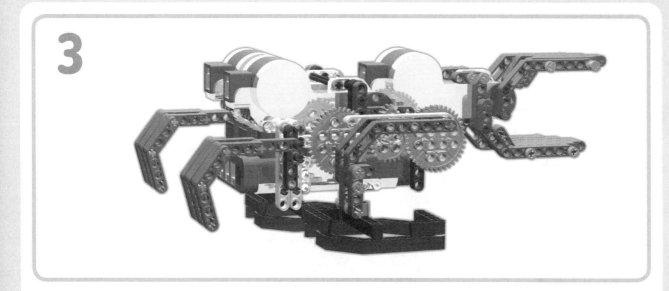

Attach the left and right front legs to the gears on their designated sides.

Place the legs in opposite positions on each side. If one leg is at the bottom, the other leg should be at the top. If one is at the left, the other side should be positioned on the right. Figures 6-4 and 6-5 show what I mean.

Figure 6-4: On the right side, place the leg pins as shown.

Figure 6-5: On the left side, place the leg pins as shown.

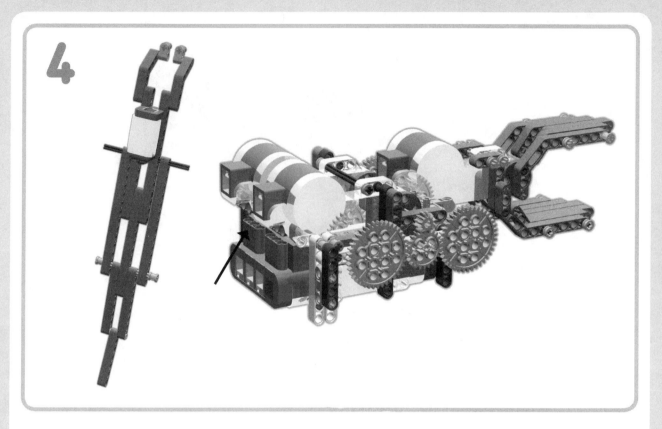

Connect the two blue pins on the tail to the rear of the alligator; then go on to steps 5 and 6 to complete the tail.

wiring connections

Connect the leg motors to ports B and C. Connect the mouth motor to port A. Connect the Touch Sensor to port 1.

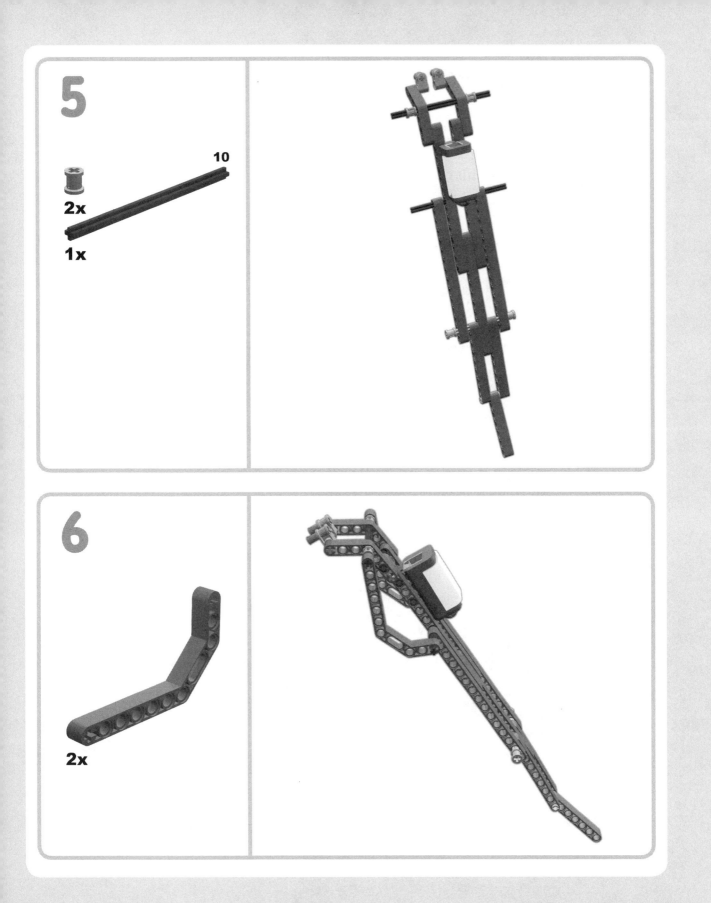

programming snout

We'll start out by creating two My Blocks—one to control Snout's legs and another to control his jaws.

my block #1: gator_walk

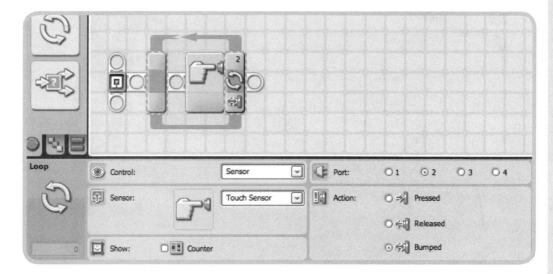

Figure 6-6: Place a Loop block on the beam. We want this loop to sense touch, so make the settings the same as the ones you see here.

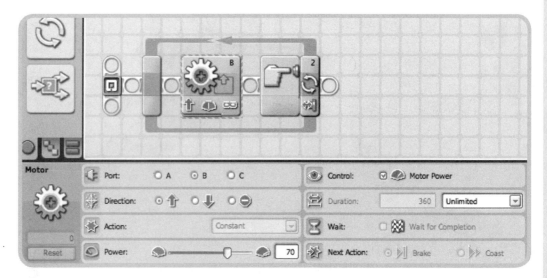

Figure 6-7: Place a Motor block in the loop. As you can see here, we're choosing port B and the up arrow. I don't know how strong your battery is, so for now start with a power setting of 70. You may need to change this setting after you test your robot.

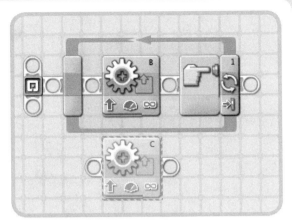

Figure 6-8: We want to place a second Motor block inside the loop below Motor block B. But when you try to insert the block, it is either forced to the side of Motor block B like this . . .

Figure 6-9: . . .or it is forced below the loop, like this.

So, here's a trick (sometimes referred to as a crowbar) to get the new block into the correct position.

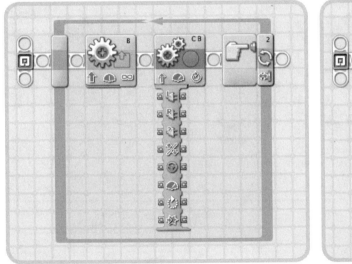

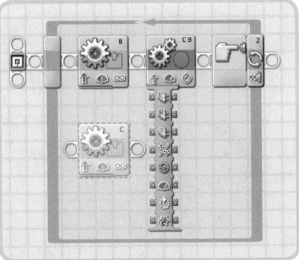

Figure 6-10: Place a Move block on the beam inside the loop, to the right of Motor block B. Expand the data hub by clicking the lower-left edge of the block.

Figure 6-11: Drag a Motor block inside the loop and place it below Motor block B. Select port C in the configuration panel so you can identify it as Motor block C.

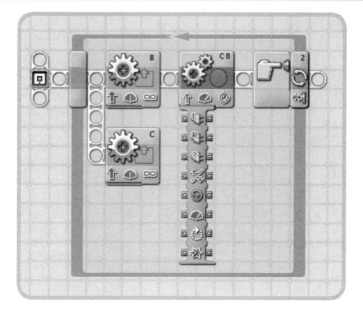

Figure 6-12: Press the SHIFT *key and drag a beam from in front of Motor block B down to Motor block C.*

Loop
breaks
with touch

motors B & C
move forward

Motor			
Port:	○ A	○ B	● C
Direction:	● ⬆	○ ⬇	○ ⊖
Action:	Constant		
Power:	━━━━⬤━━━	70	
Control:	☑ Motor Power		
Duration:	360	Unlimited	
Wait:	☐ Wait for Completion		
Next Action:	● Brake	○ Coast	

Reset

Figure 6-13: Delete the Move block and configure Motor block C as shown here.

Now, select the whole loop with your mouse and click the Create My Block button on the Toolbar at the top (the two small blue bars). Name the block *gator_walk* and give it any symbol you like.

my block #2: gator_mouth

The next My Block is a bit more complicated. This block will tell the mouth to open and close. I'll also use it to make the robot respond better when I want it to stop.

BEWARE: UNRESPONSIVE LOOPS

One of the problems with Touch loops is that they recognize the touch at only one specific point in the action—right before it starts a new loop. Figure 6-14 shows an example.

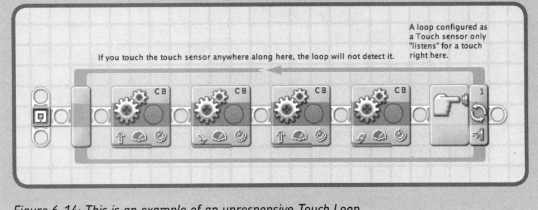

Figure 6-14: This is an example of an unresponsive Touch Loop.

There are many ways to avoid the problem of an unresponsive loop sensor. In this program I'll add extra sensors.

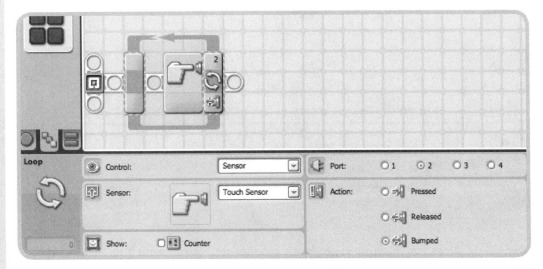

Figure 6-15: In a new window, place a Loop block on the beam. As seen here, this block will be configured as a Touch Sensor.

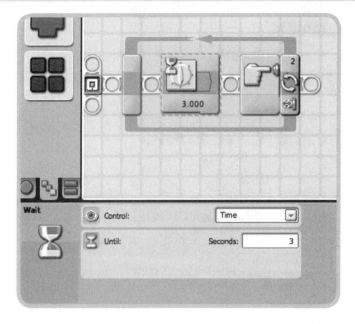

Figure 6-16: Because we want the jaws to stay closed for 3 seconds between openings, we'll place a Wait Sensor in the loop. Configure it as a Time Sensor and type 3 seconds (as shown here).

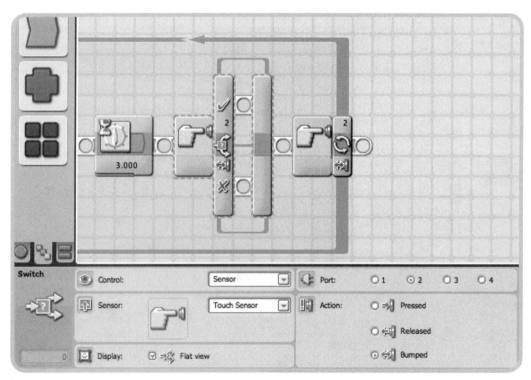

Figure 6-17: As part of my strategy to get my robot to stop when I want it to, we'll put a switch in the loop. Make this Switch block a Touch Sensor. This Switch block will say, "If the sensor has been touched, stop moving; otherwise, move to the next step."

In a switch, if the response to the setting is true (something has touched the sensor!), the program jumps to the top beam on the switch. If the answer is false (no touch yet!), it takes the low road (or in this case, the lower beam).

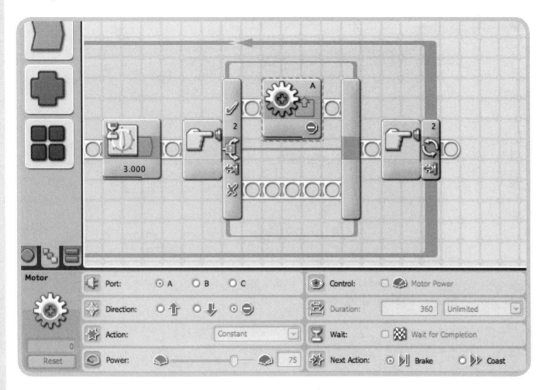

Figure 6-18: To make this happen, place a Motor block on the top beam (inside the Switch block) and configure it to stop the motor, as shown here.

If the switch statement is false (no button pushed), we want the robot to move right on to the next block (outside the Switch block). So we'll just leave that lower beam empty.

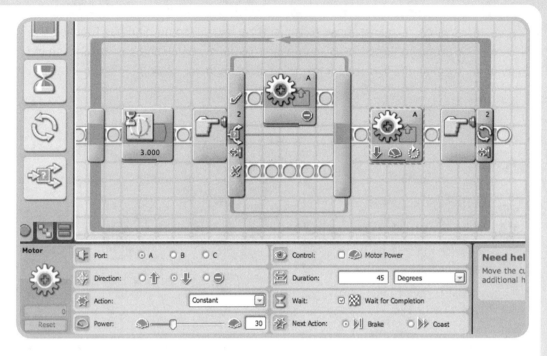

Figure 6-19: After the switch, we'll place a Motor block on the beam and configure it to open Snout's mouth. Use the settings shown here.

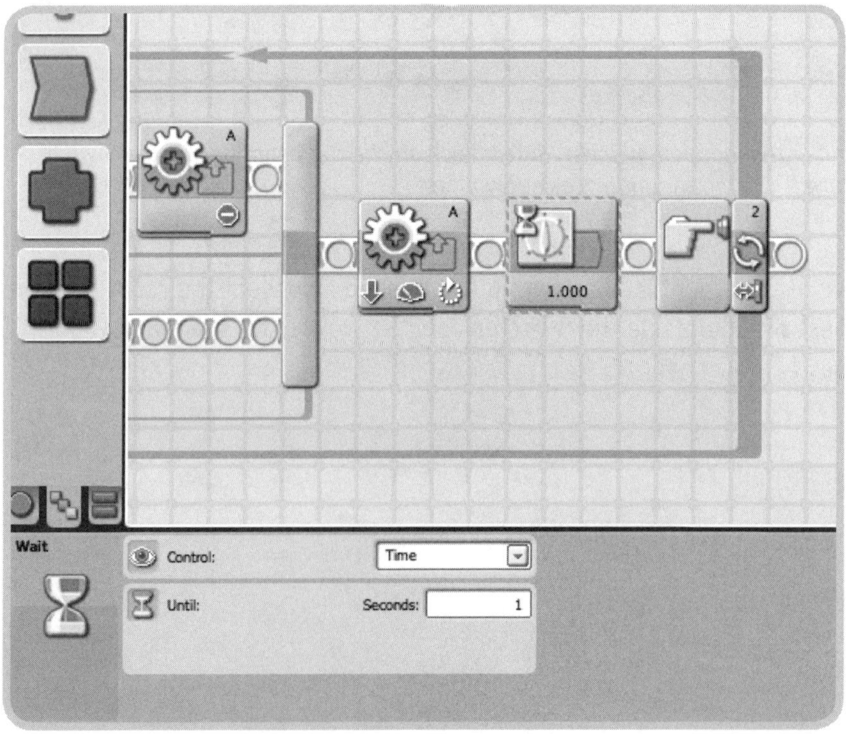

Figure 6-20: The next block, a Wait block, tells Snout to keep its mouth open for one second.

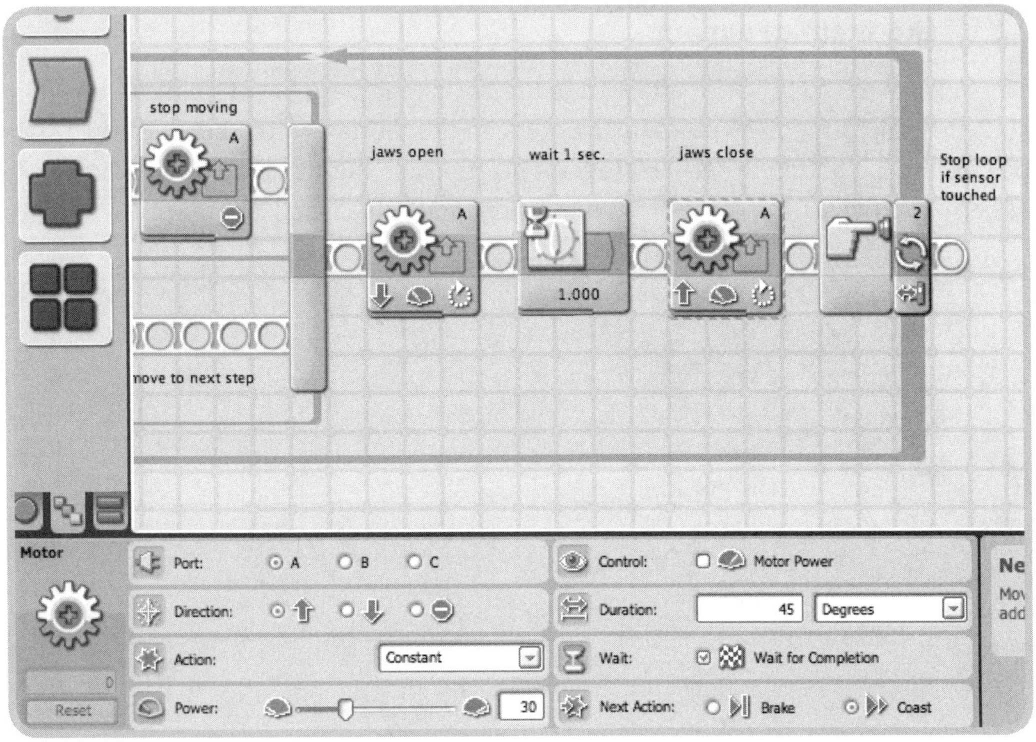

Figure 6-21: Now add another Motor block to shut the jaws. The settings are shown here.

Now we have a little program that checks for a touch before the alligator's mouth opens and after it closes. The following figure shows the full code for this My Block.

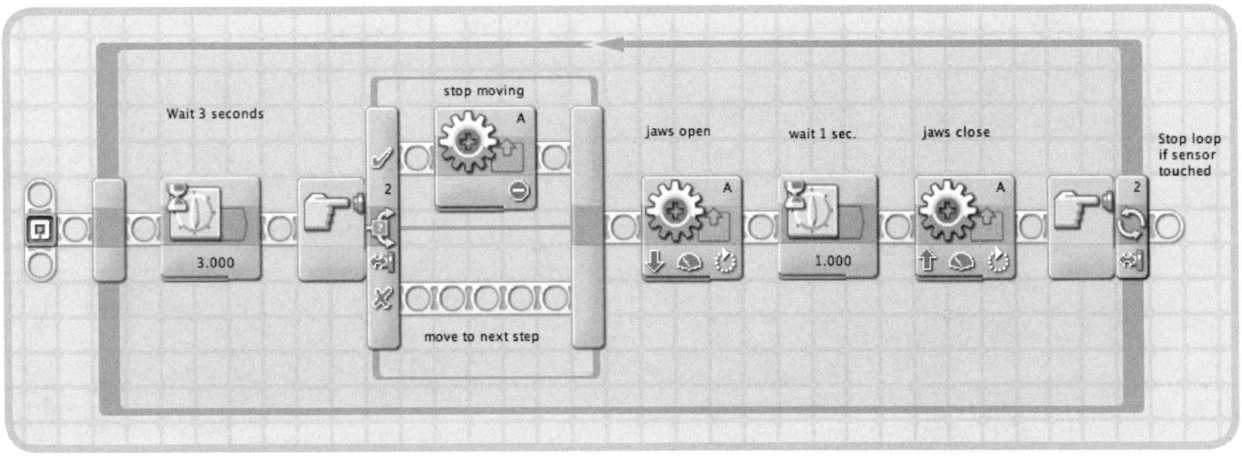

Figure 6-22: The complete gator_mouth My Block

Select the whole loop and save it as a My Block named *gator_mouth*. (See page 11 if you need some help creating a My Block.)

the main program

Since we have the NXT brick attached to the bottom of the robot, it's difficult to start the robot's action by pressing the orange Enter button on the brick. You will still use the orange button to set the program to "running", but this bit of programming will allow you to set the robot on its feet before you actually make it move.

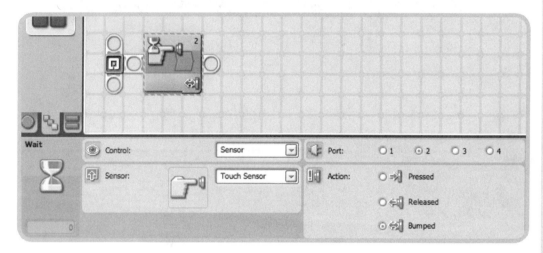

Figure 6-23: The first block in the main program is a Wait Touch Sensor. Place it on the beam and enter the settings shown here.

To be sure the robot is ready to go, I'll have Snout say, "Watch out!" when I touch the sensor the first time.

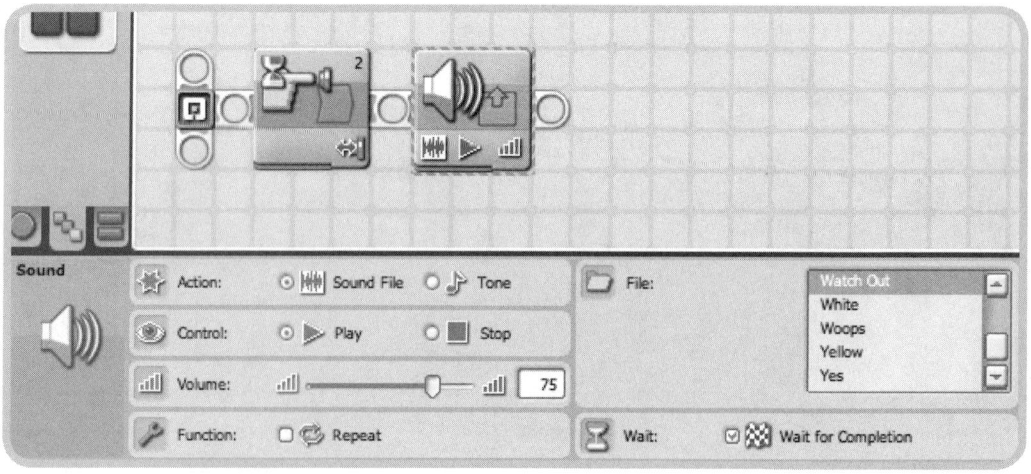

Figure 6-24: I do this by inserting a sound block configured as shown here (you can choose another sound if you like). In most cases, you'll want the volume set at 100 because sound files tend to be very quiet. In this case, the message is for my ears only, so I want it to be quiet.

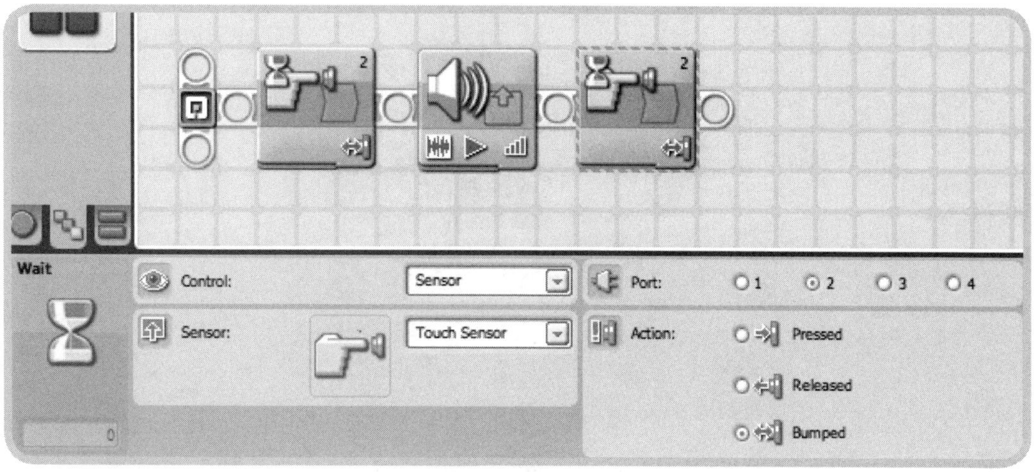

Figure 6-25: Next we'll place another Wait Touch block. This means that after we hear "Watch out!" (or whatever sound was chosen), we'll press the Touch Sensor again and Snout will go into action.

For the last part of the program we'll run two My Blocks at the same time on two parallel beams.

Click the blue bars at the bottom left of the screen. (This is the Custom palette.) Two turquoise symbols will appear at the top left. Click the top symbol, and you'll see all the My Blocks we created. (You'll see the My Block names as you mouse over the blocks.)

Figure 6-26: Choose the gator_walk block and drag it to the beam after the Touch Sensor, as shown here.

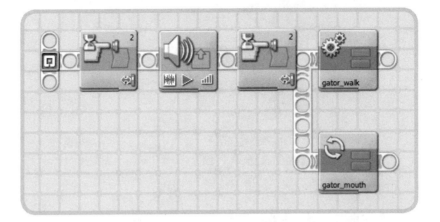

Figure 6-27: Now click the My Block symbol again and select the gator_mouth block. Drag it to a spot below the gator_walk block. Now press the SHIFT key and use your mouse to drag a connection from the beam above to the gator_mouth block. We want both of these My Blocks to run at the same time.

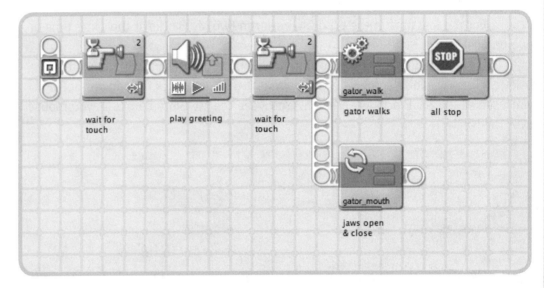

Figure 6-28: Finally, to stop this robot with the third Sensor Touch, I'm ending the program with a Stop block, which stops all action.

That's it! You're ready to download the program and watch your alligator go.

NOTE Before you test the program, make sure that your sensors are hooked up to the correct ports. It's less confusing if you use ports B and C when you are using two motors. Use port A when using only one motor in a program. If you do the same thing every time, you'll make fewer mistakes.

LEGOsaurus: an NXT dinosaur

My original inspiration for building this LEGOsaurus was the cool orange TECHNIC spikes (the 1 × 3 BIONICLE tooth). You may not have as many spikes in your parts stash as this robot needs, but don't let that stop you. You can make your own spikes from foam board or cardboard—just punch holes in them so you can attach them to the axle pins.

   The gear arrangement on the base of this robot is one you can use for many different four-legged animals. But be careful as you build it, because the NXT brick is attached underneath the robot, facing down. That means that there is a good chance you will accidentally turn the robot on during construction. To save battery strength, be sure to check periodically that the brick has not turned on.

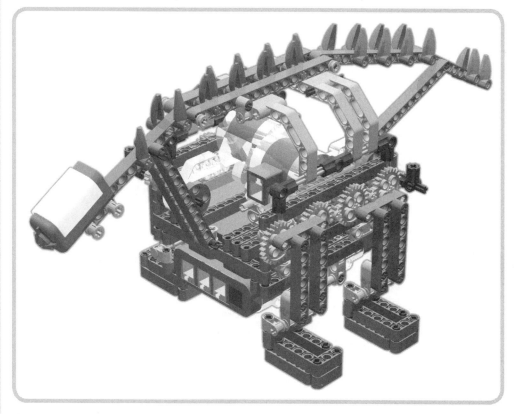

Figure 7-1: A plodding NXT stegosaurus

building LEGOsaurus

LEGOsaurus conveniently turns on and off with a Touch Sensor.

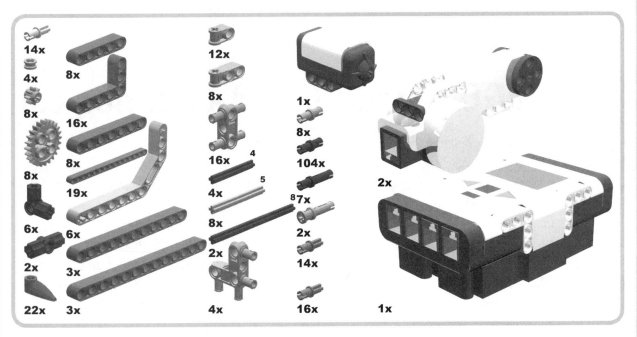

base

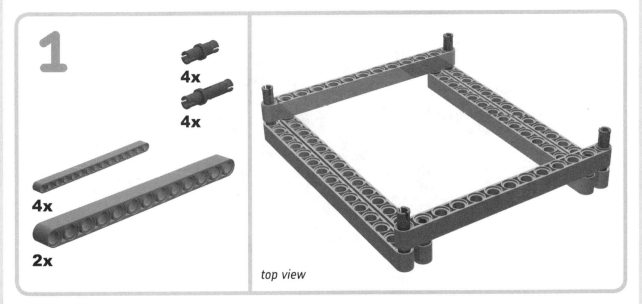

top view

2

4x

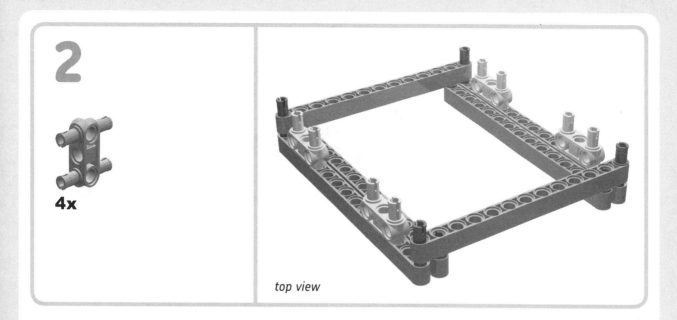

top view

PART SUBSTITUTION SUGGESTION

In this step, two long pins plus one 3-hole beam (Figure 7-2) can substitute one of these pieces (Figure 7-3).

Figure 7-2: You can use this combination of pieces . . .

Figure 7-3: . . . in place of this piece.

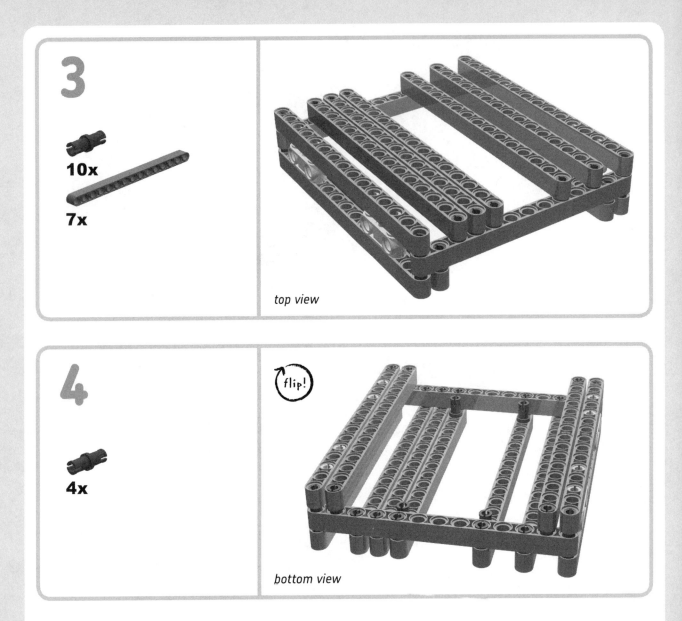

3

10x

7x

top view

4

flip!

4x

bottom view

These four pins will help attach the brick to the bottom of the robot.

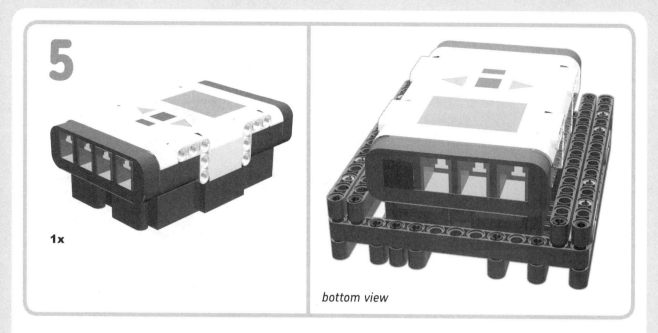

5

1x

bottom view

Snap the brick onto the four pins from step 4. Be sure to hold the brick carefully as you attach it in the next two steps.

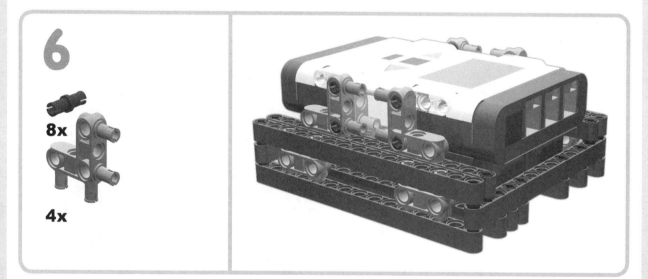

6

8x

4x

You need to lift the brick out to attach these connectors.

7

14x

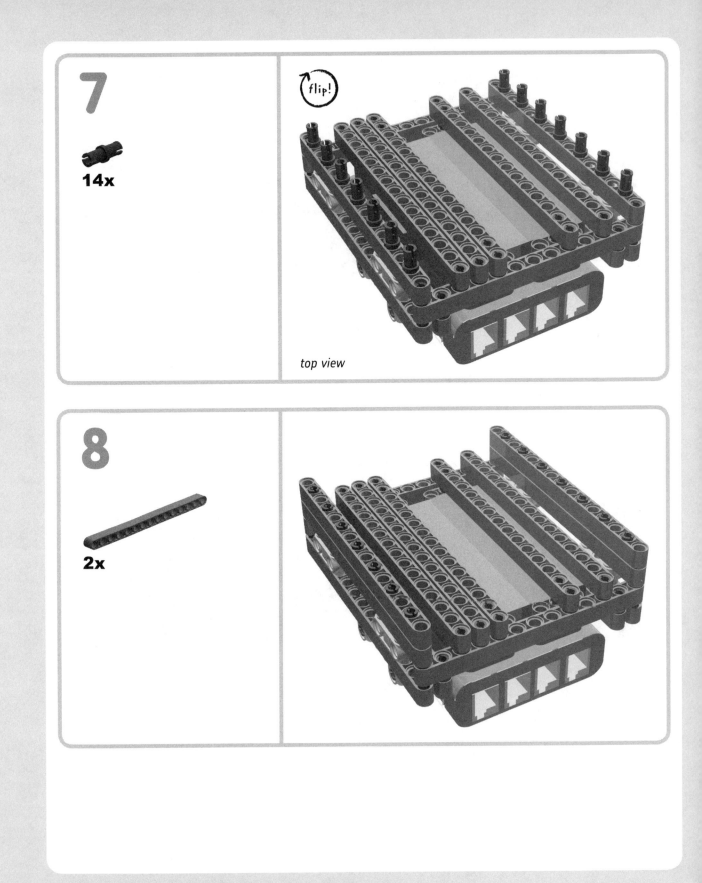

flip!

top view

8

2x

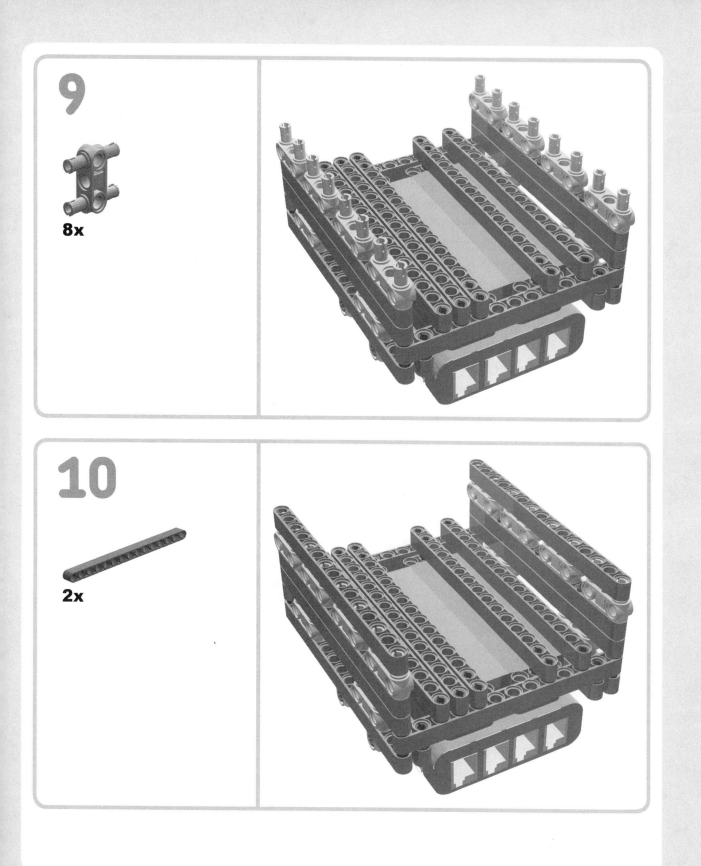

9

8x

10

2x

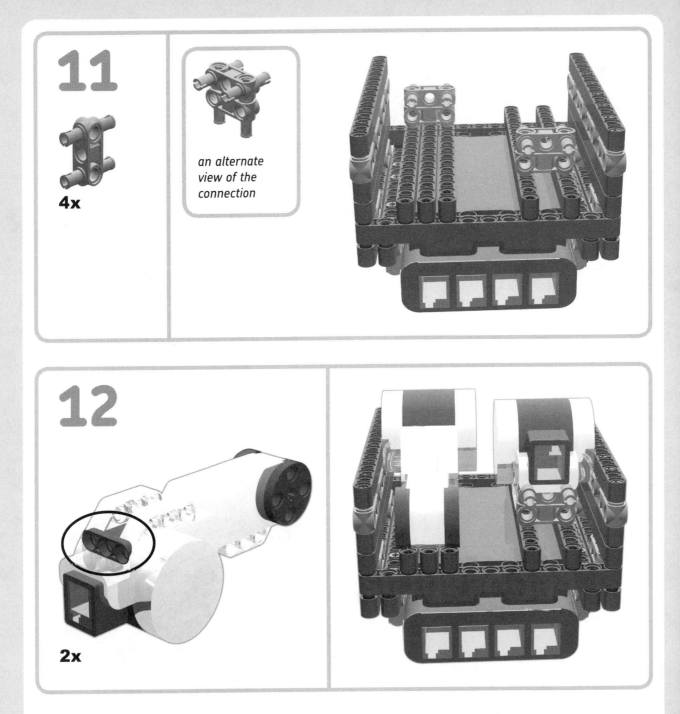

11 4x

an alternate view of the connection

12 2x

The pins fit very easily into these holes at the end of the motor.

legs (build four)

While I think the legs and feet are just right on this robot, it's possible that you might have an equally good idea for constructing similar feet with the parts you have. Just don't make the feet too big.

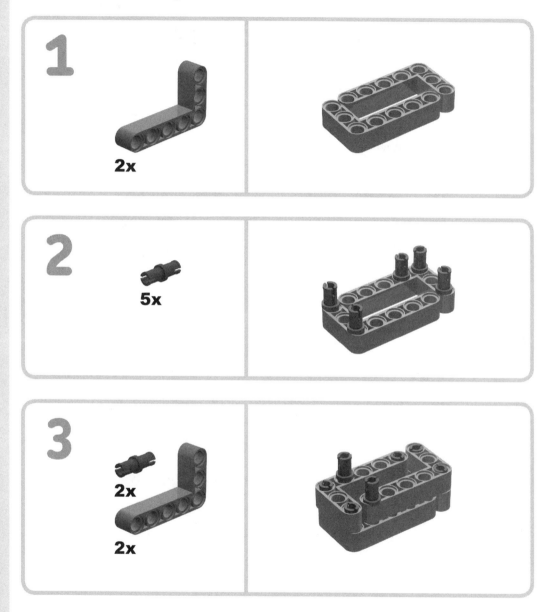

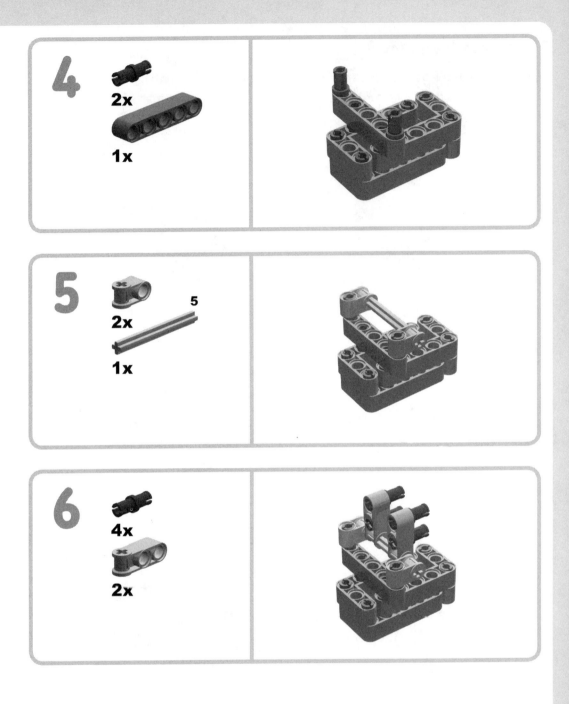

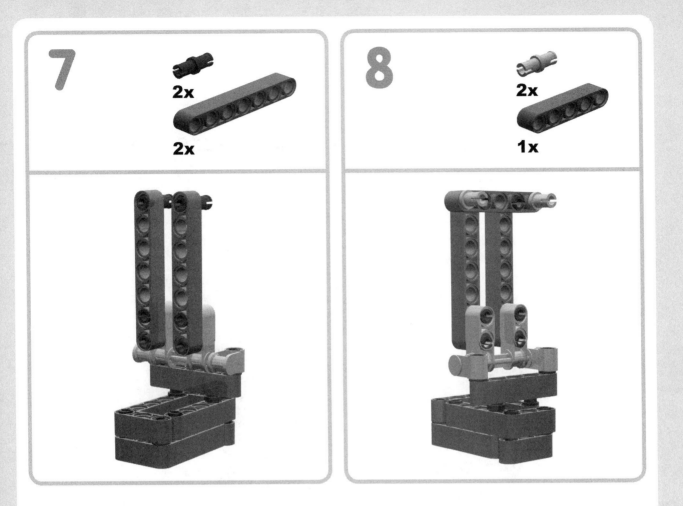

Use gray pins in step 8.

armor

The beams that make up the backbone of the dinosaur (and that hold the spikes) could be assembled out of different sized beams. It just takes a bit of experimentation.

PART SUBSTITUTION SUGGESTION

If you don't have enough orange BIONICLE teeth, you can create spikes with cardboard or foam sheets. (The alternate style spikes in Figure 7-4 are created with a 79-cent foam sheet, which can be found in any craft store.) Just cut out any shape you like (I chose a kite shape), trace it on the foam, cut it out, and punch a hole for a pin. Use black pins to connect the spikes to the beams. A small hole punch or the point of a protractor will make a hole just large enough for a black pin. (A regular hole punch makes too large a hole.) It will be more difficult to force the pin through the small hole, but it results in a very firm connection.

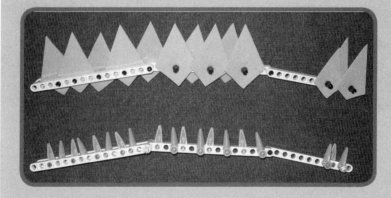

Figure 7-4: Making spikes from foam sheets

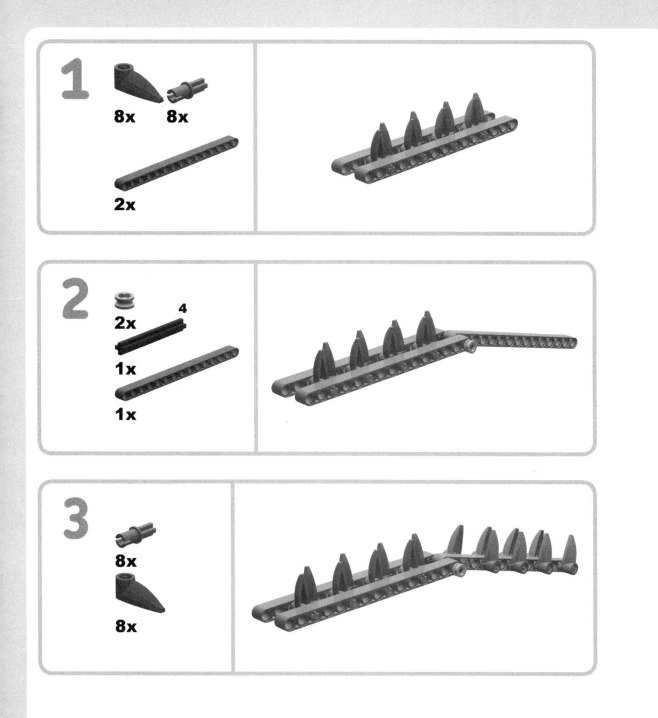

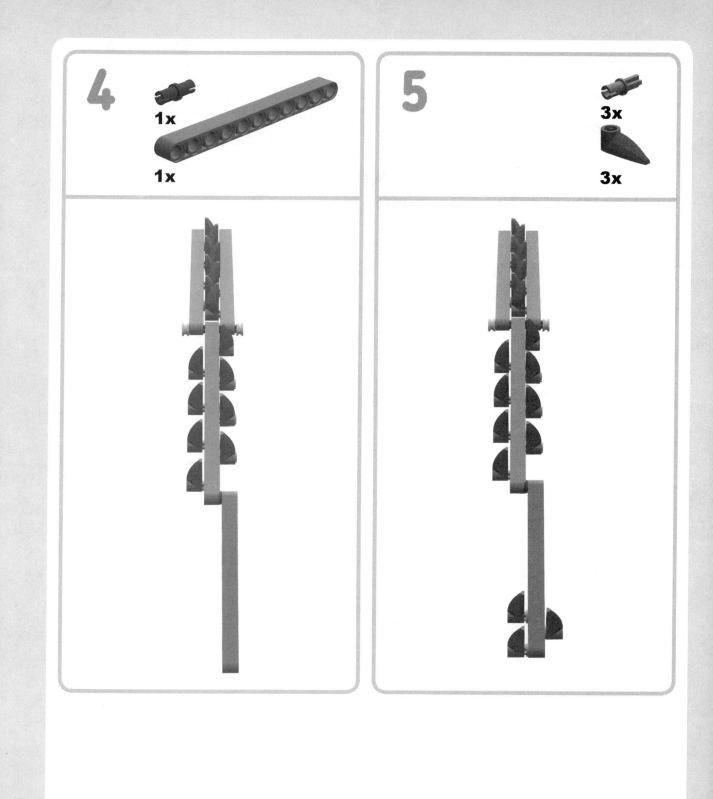

full assembly

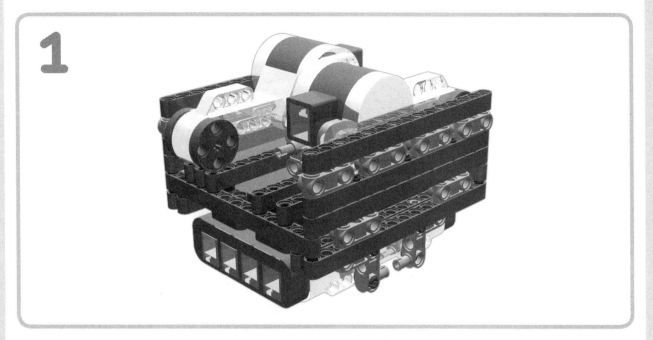

Start with the base you already created. This end of the robot (on the left in this view) will be the tail end, which is the end with ports 1, 2, 3, and 4.

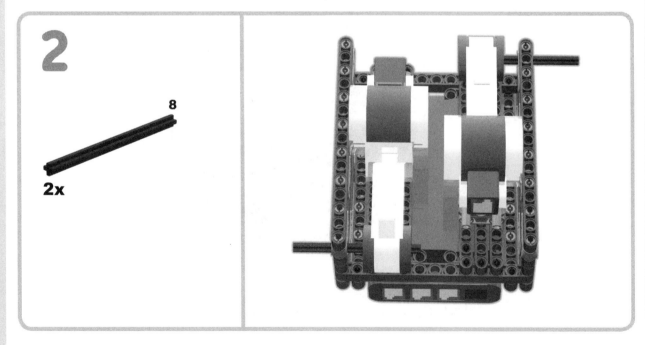

8

2x

Insert #8 axles into the motors. These axles will drive the four legs. The side of the brick with three ports (A, B, and C) is the front end.

3

14x

Insert seven yellow axle pins on each side (axle end out).

4

8x

8x

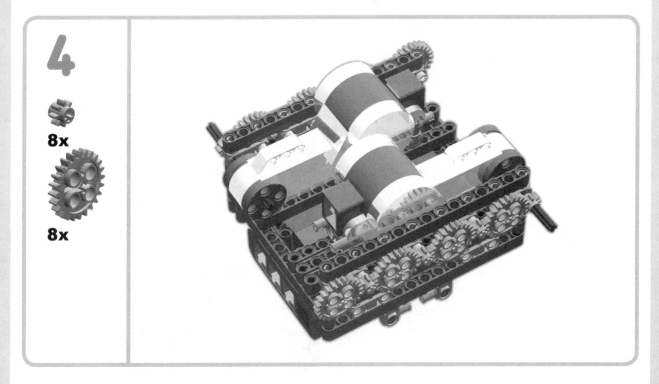

Place the gears on the yellow pins, as shown.

5

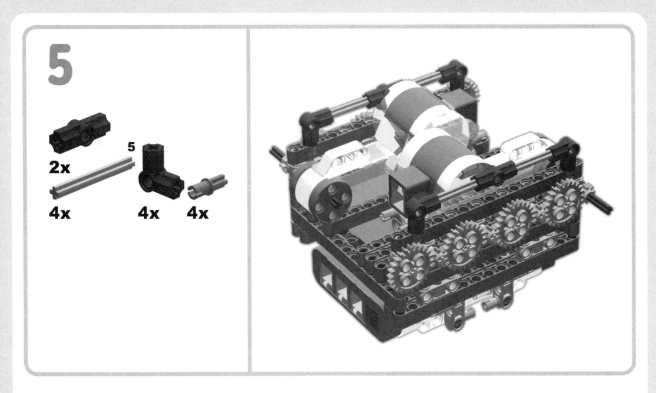

2x **4x** 5 **4x** **4x**

Use blue axle pins here. This step is designed to show you where these parts will go. You'll install them with the beams in step 6.

6

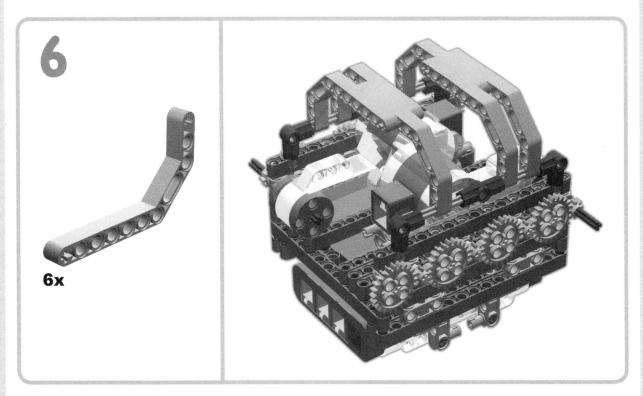

6x

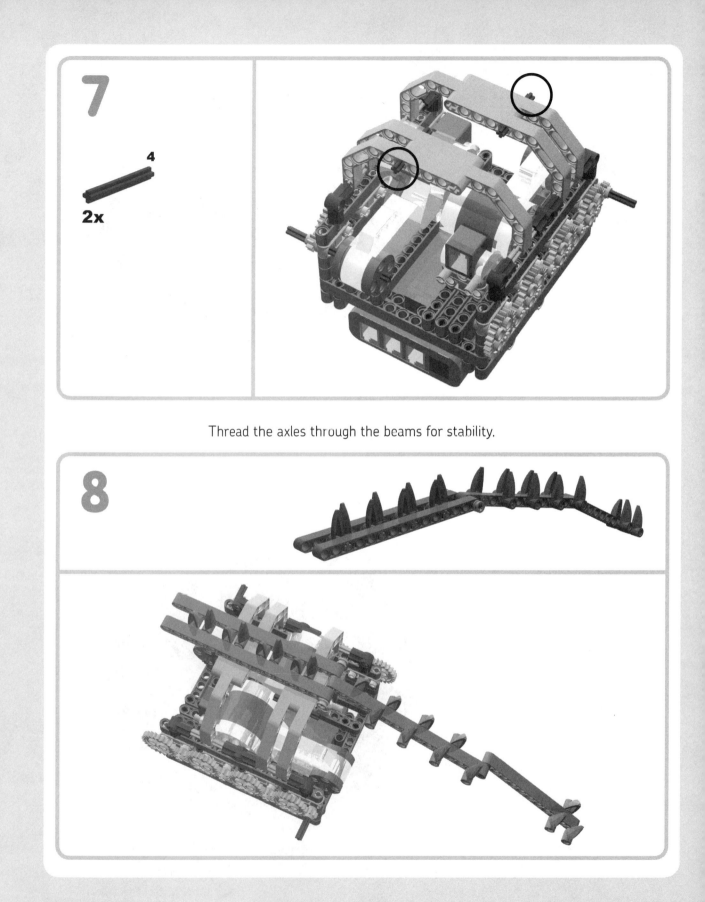

Thread the axles through the beams for stability.

Insert the blue axle pin (axle-side first) into the connector. Then insert the black TECHNIC pin into the other connector hole. Use these pins to attach the scales to the gray ribs in two places (wherever they fit with the least stress).

10

1x

1x

1x

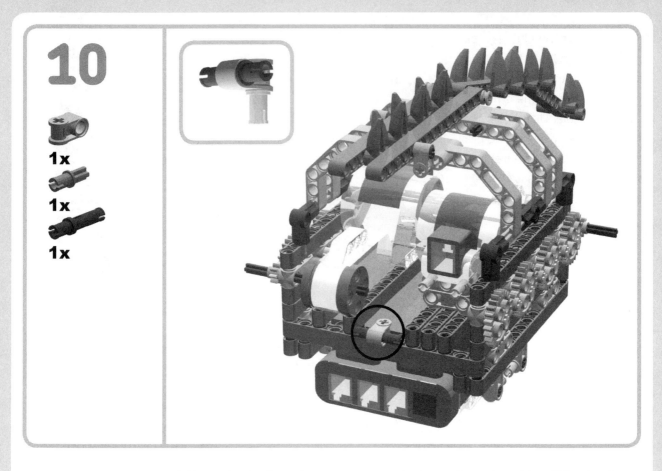

A blue axle pin fixes the connector to the base. Note that this will be the front of the dinosaur.

11

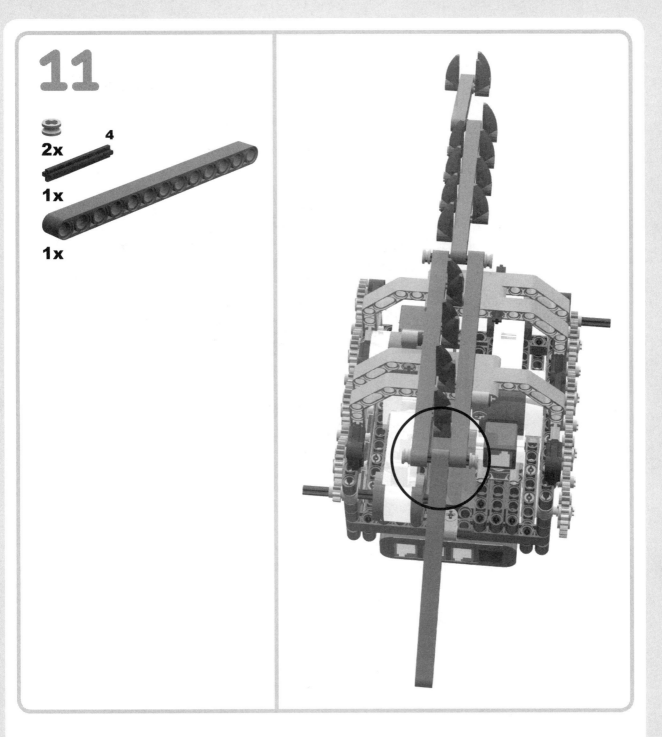

Connect the 13-hole beam to the front end with the axle and bushings.

12

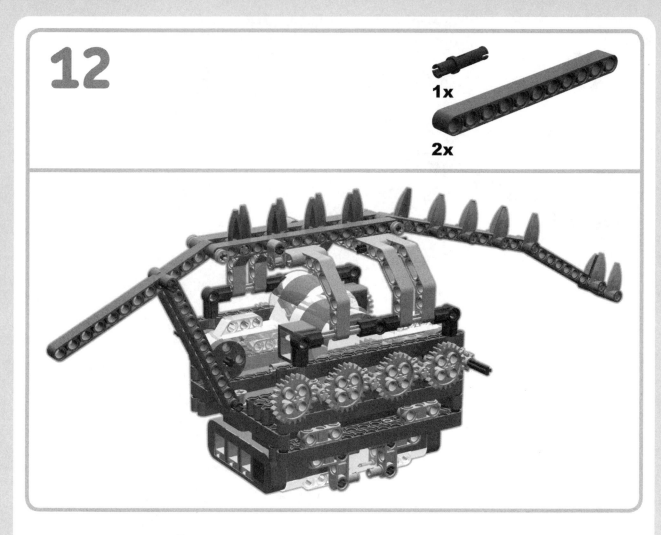

The 11-hole beams are attached to the connector piece you constructed in step 10. A long black pin holds them where they meet the 13-hole beam.

13

1x

1x

1x

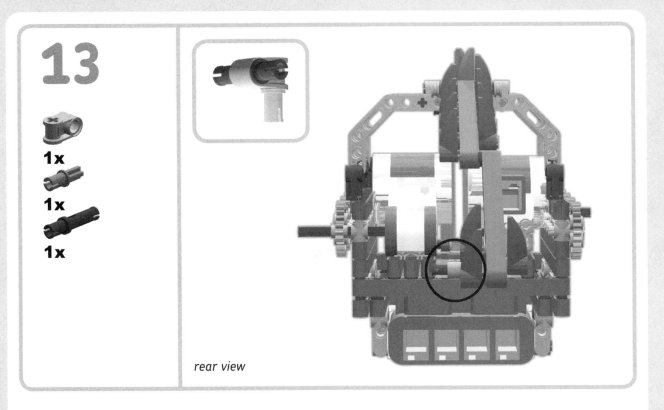

rear view

Step 14 shows another view of where these parts go.

14

1x

1x

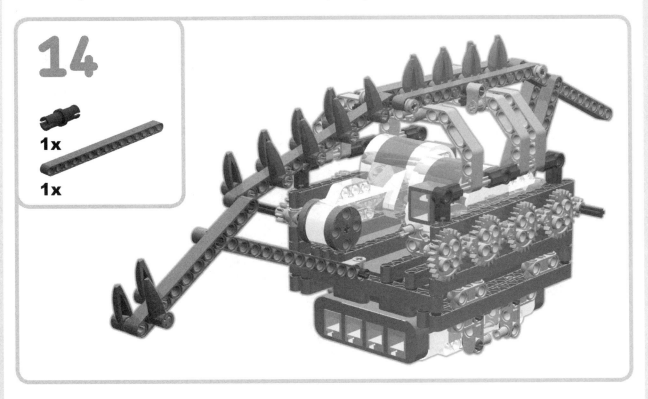

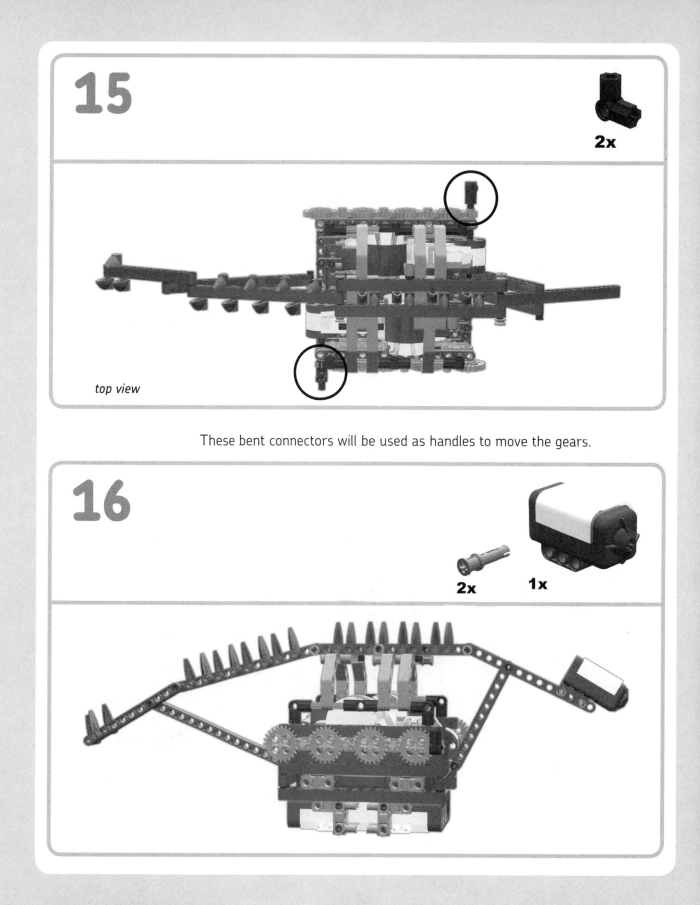

15

2x

top view

These bent connectors will be used as handles to move the gears.

16

2x 1x

17

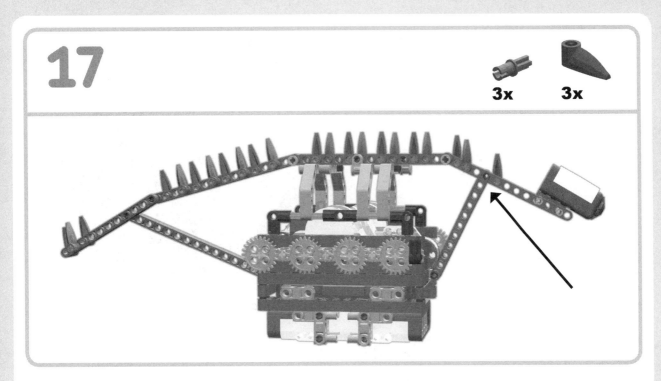

Use blue axle pins and orange teeth (or your alternate version) to complete the spikes.

18

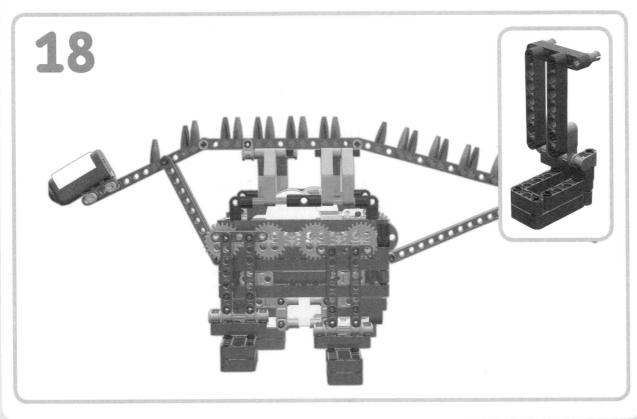

It is important that the gears on each side are lined up exactly the same way—particularly for the two gears supporting each leg. Ideally, as you face each side of the robot, the gears and legs should be in the position shown in Figure 7-5 when you are ready to run it.

Figure 7-5: Correct positioning for LEGOsaurus's gears and legs

wiring connections

Using the shortest possible electric cable, connect the motors to ports B and C. Connect the Touch Sensor to port 2. Be sure to position the wires so they will not interfere with the gears or legs.

need some help troubleshooting this robot?

You'll find tips for troubleshooting walking robots in Appendix B. Keep in mind that these robots have been tested thoroughly—with both experienced and inexperienced builders—so there *is* a solution if your robot isn't walking.

programming LEGOsaurus

As we did with other robots that have the NXT brick underneath them, we'll include a little startup routine, so you don't have to start the robot before you set it down.

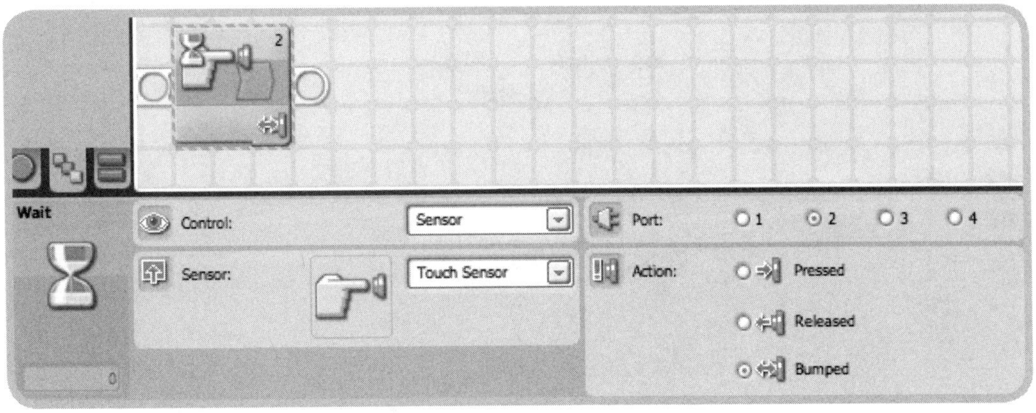

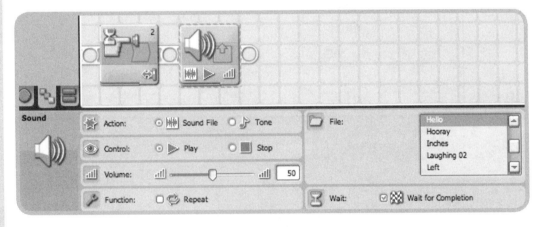

Figure 7-6: This program begins with the Wait block configured as a Touch Sensor. Choose the settings shown here.

Figure 7-7: Next, add a Sound block. From among the options in the File, I chose Hello, so I know my dinosaur will move when I bump (press and release) the Touch Sensor again. Notice that I set the sound level to 50. This is because the sound is only for my benefit; I don't want it to be heard by the rest of the world.

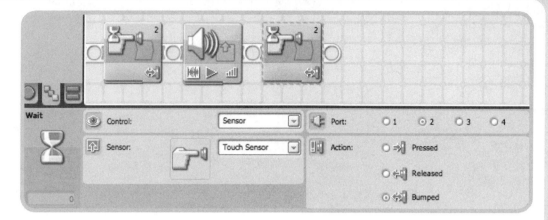

Figure 7-8: The second Wait block is configured just like the first one.

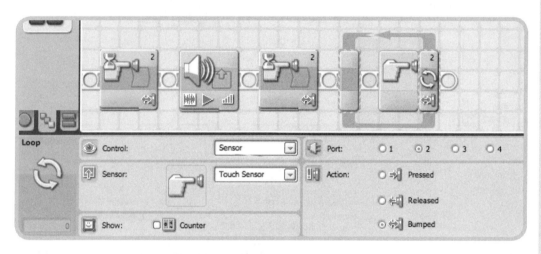

Figure 7-9: Now we'll drop a Loop block onto the beam. I configured the loop to end when the Touch Sensor is bumped. The settings for the loop are shown here.

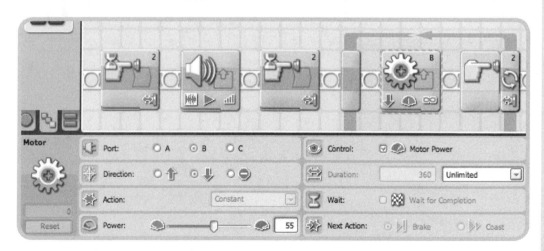

Figure 7-10: Next, place a Motor block inside the loop. Use the settings shown here.

We need to expand the loop temporarily to insert a parallel beam inside it.

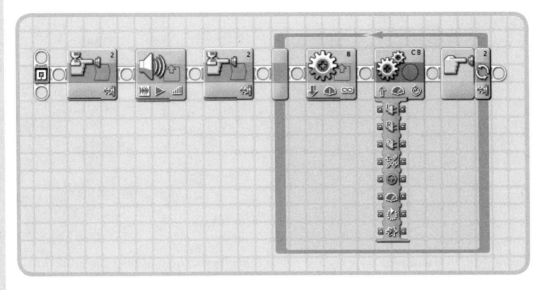

Figure 7-11: Place a Move block inside the loop after the Motor block. Expand its data hub by clicking your cursor on the bottom-left corner of the Move block.

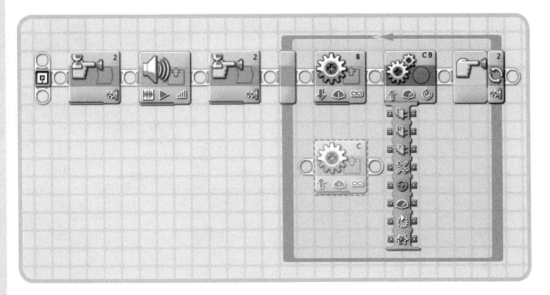

Figure 7-12: Now you have a space to place another Motor block beneath the B Motor block. For this new block, select port C in the configuration panel.

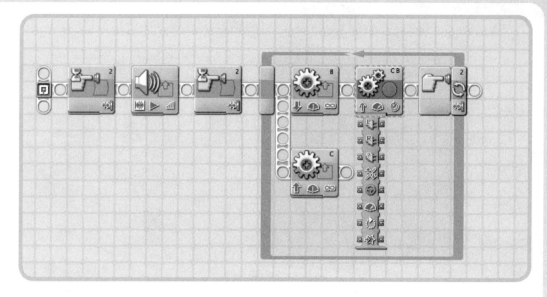

Figure 7-13: Hold down the SHIFT key and drag a beam from in front of the B Motor block down to the C Motor block.

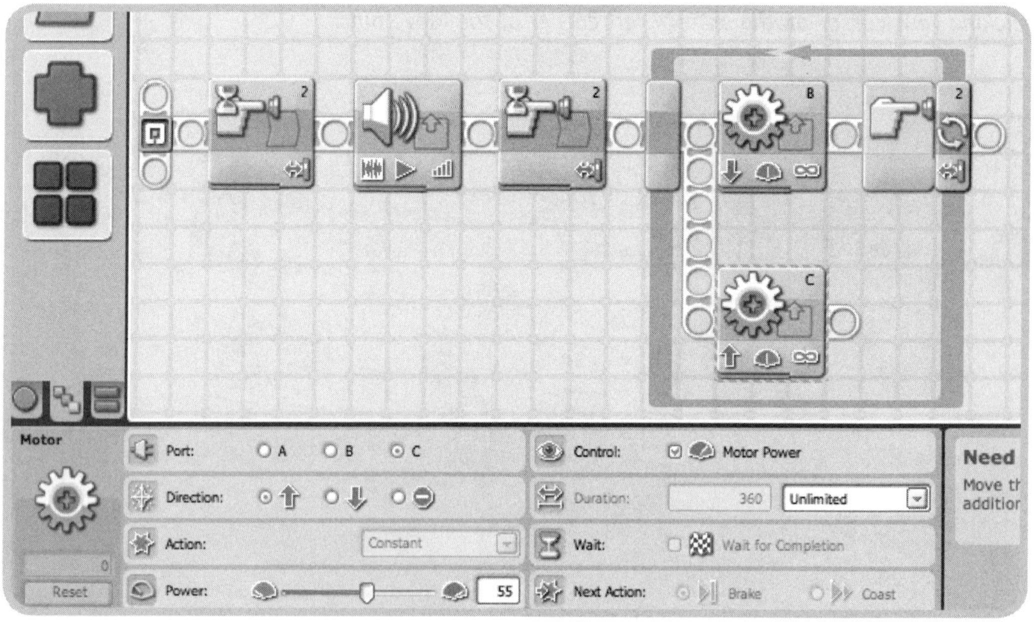

Figure 7-14: Delete the Move block used to expand the loop, and configure the C Motor block as shown here.

Once LEGOsaurus begins moving, the loop will break and move on to the next block when the Touch Sensor is bumped.

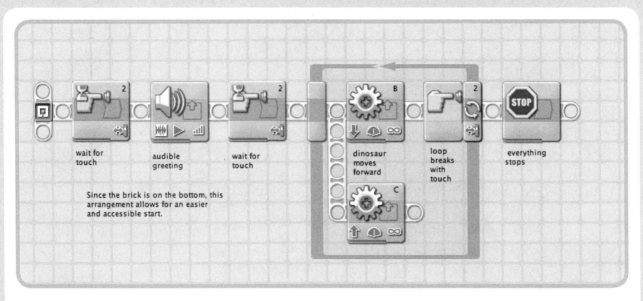

Figure 7-15: Add a Stop block after the loop, which will stop everything.

This completes the program for LEGOsaurus. Now you are ready to download the program and test drive your robot.

NOTE Remember, you start the program by pressing and releasing the sensor once. When you hear "Hello," press and release the sensor again. Your dinosaur will be off and running.

pygmy: an NXT elephant

I have to admit, I'd never heard of the pygmy elephant before I designed this model, but when I learned about it, it was easy to see that this robot was just that kind of elephant.

Originally, the design of this robot included large gears and long legs, but they did not work well—at least not well enough to be able to promise you that the robot would work (instead of having the legs randomly rip off).

This model seems particularly prone to turning instead of walking straight. The solution for that problem is to check the position of the gears on the side the robot is turning toward. Even a subtle drag of a moving part against a nonmoving part alters the function of the robot.

The trunk can actually be made out of many different things if you don't have enough of the round pieces I used. Cut up a straw, use a black marker to color uncooked ziti pasta—be creative. Also, the trunk does not have to be a particular length. Add what looks best to you. Another note about the trunk—I will suggest a few places to attach it to the elephant; but each different arrangement makes the trunk curl in different ways, so experiment to see what you prefer. (Everyone in my family tied it a different way.)

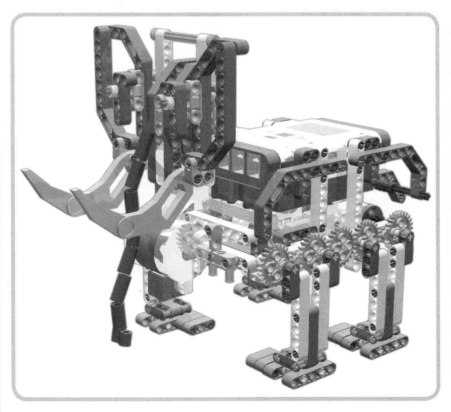

Figure 8-1: Pygmy!

building pygmy

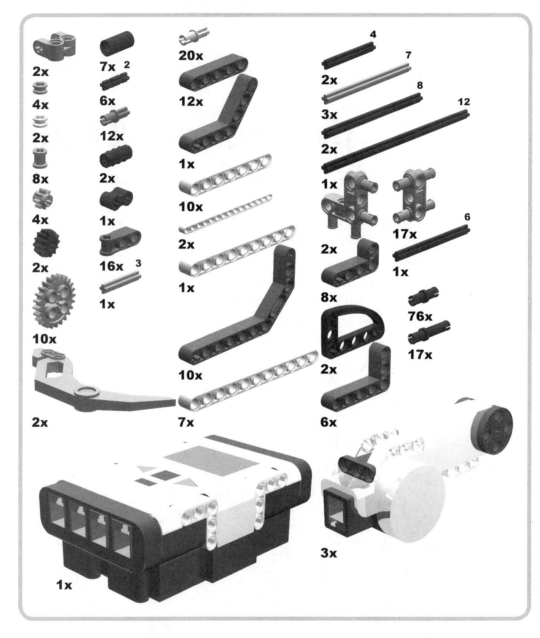

* Non-LEGO part required: Fishing line

base

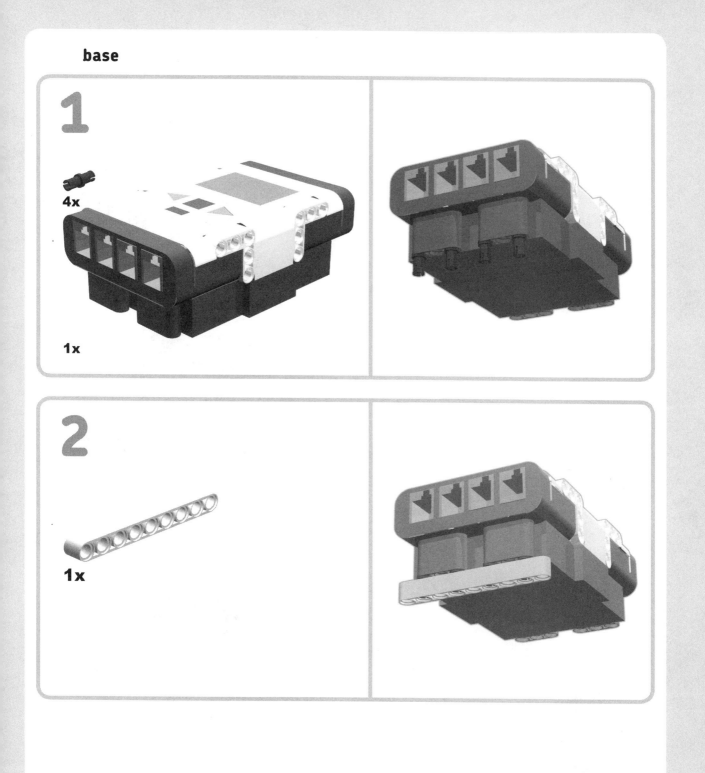

1

4x

1x

2

1x

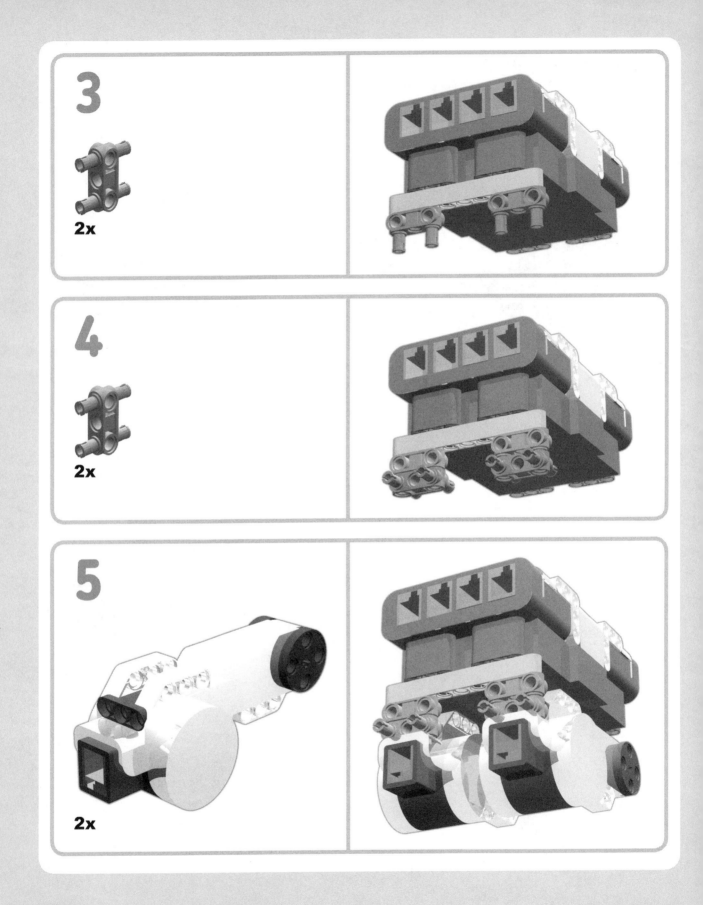

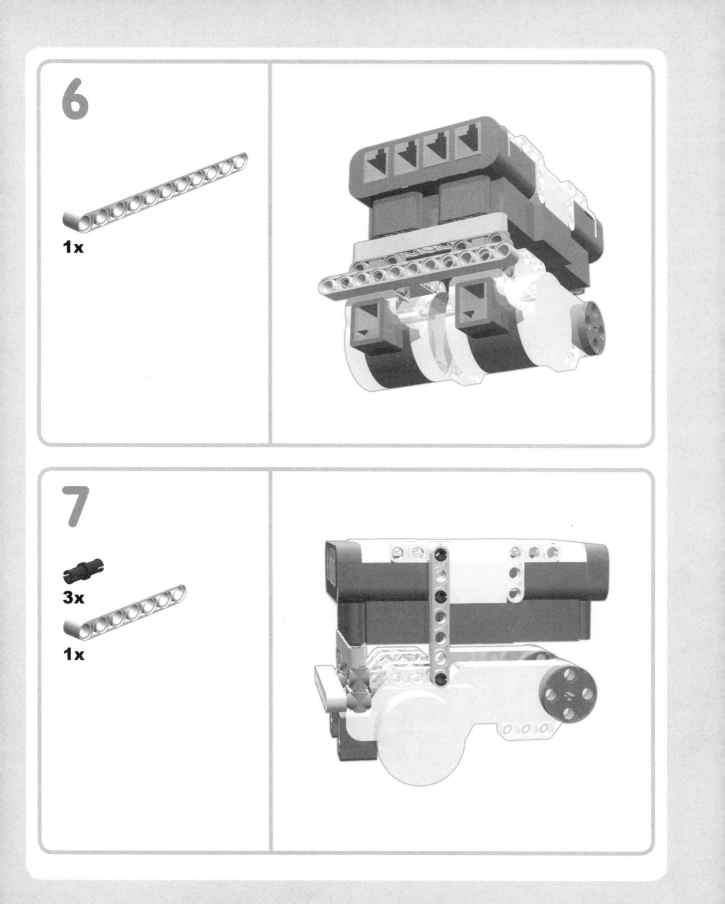

6

1x

7

3x

1x

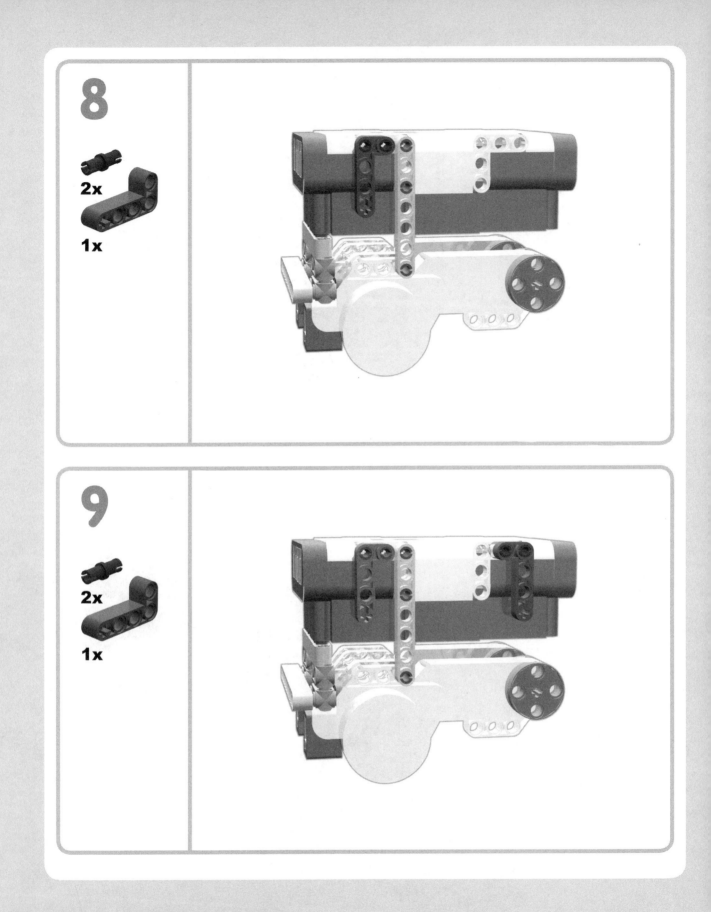

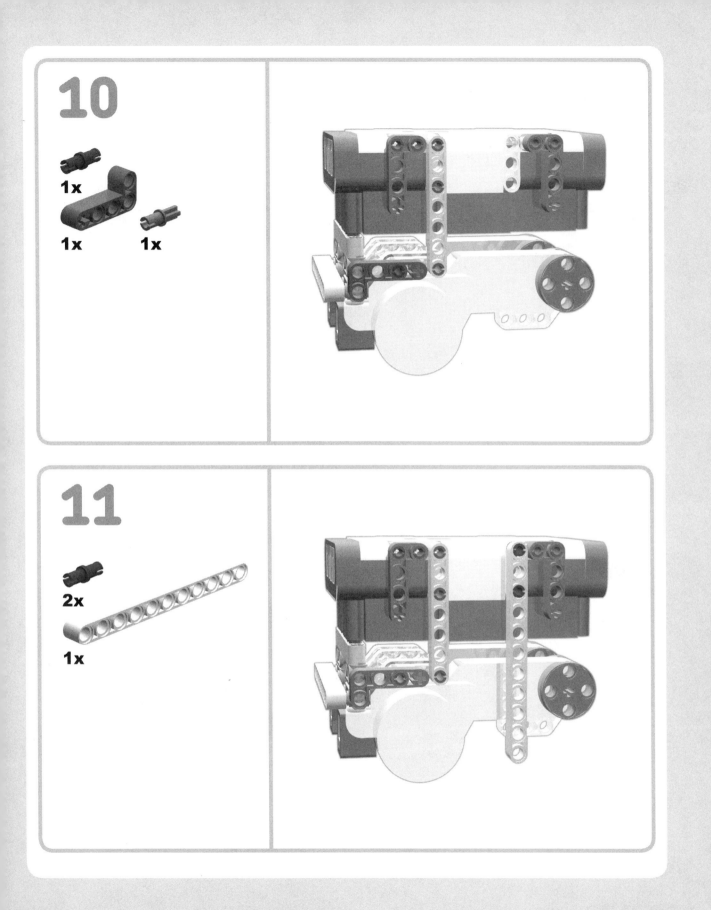

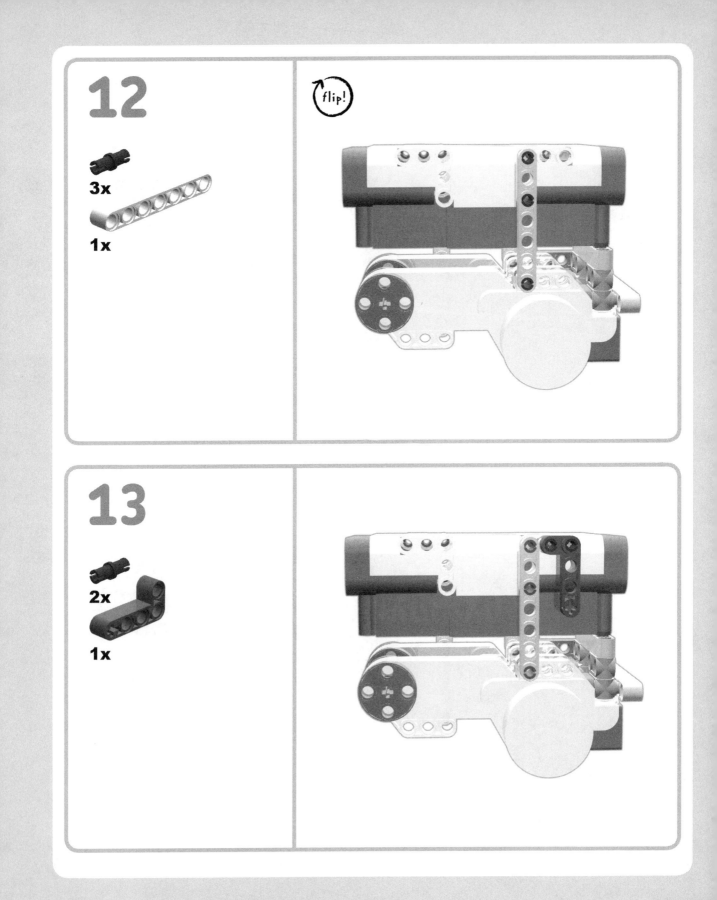

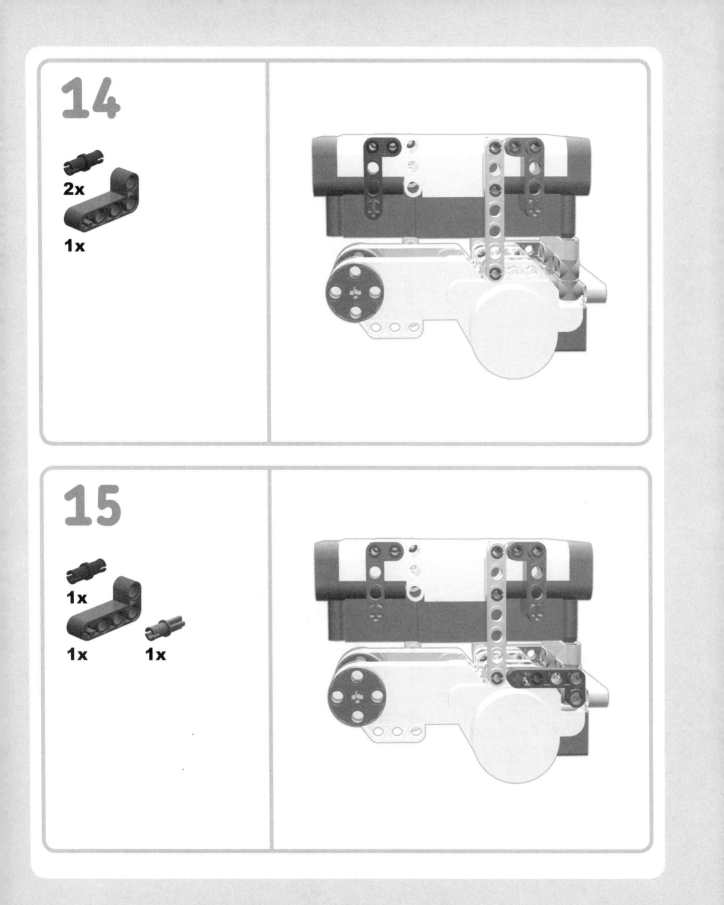

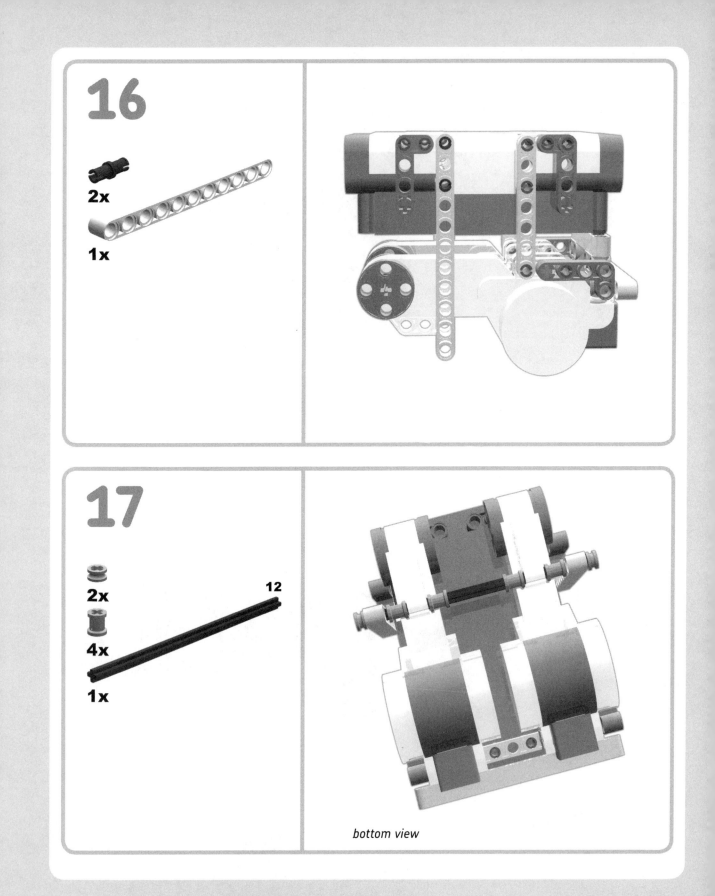

16

2x

1x

17

2x

4x

12

1x

bottom view

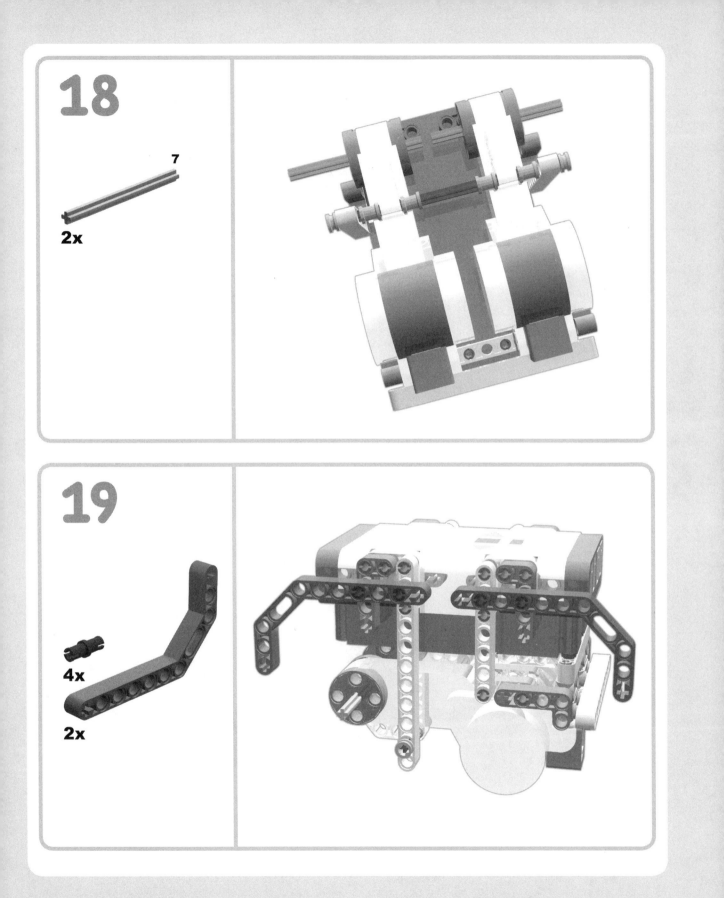

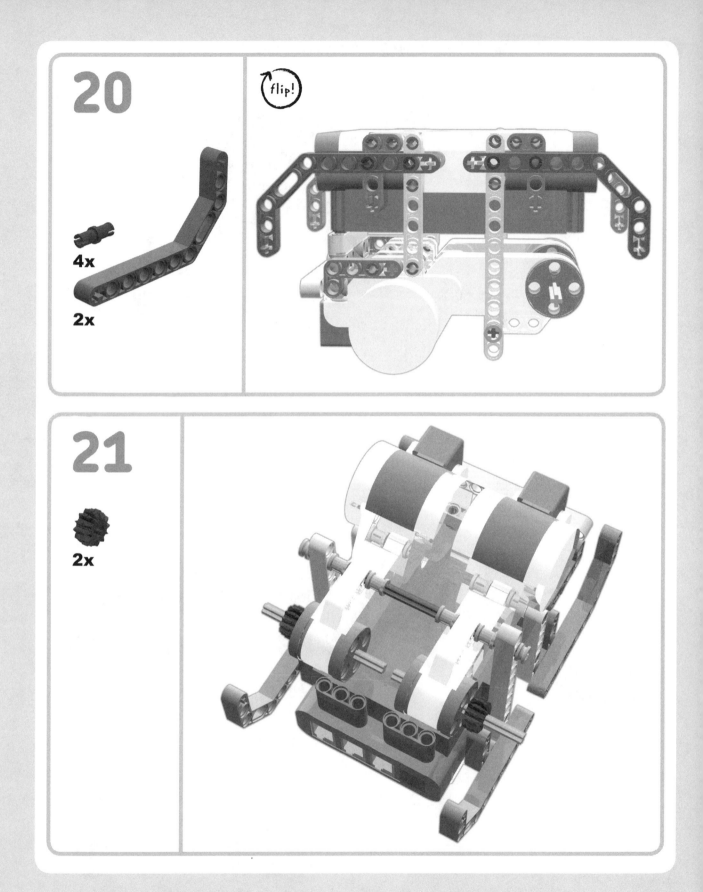

22

2x

1x

23

6x

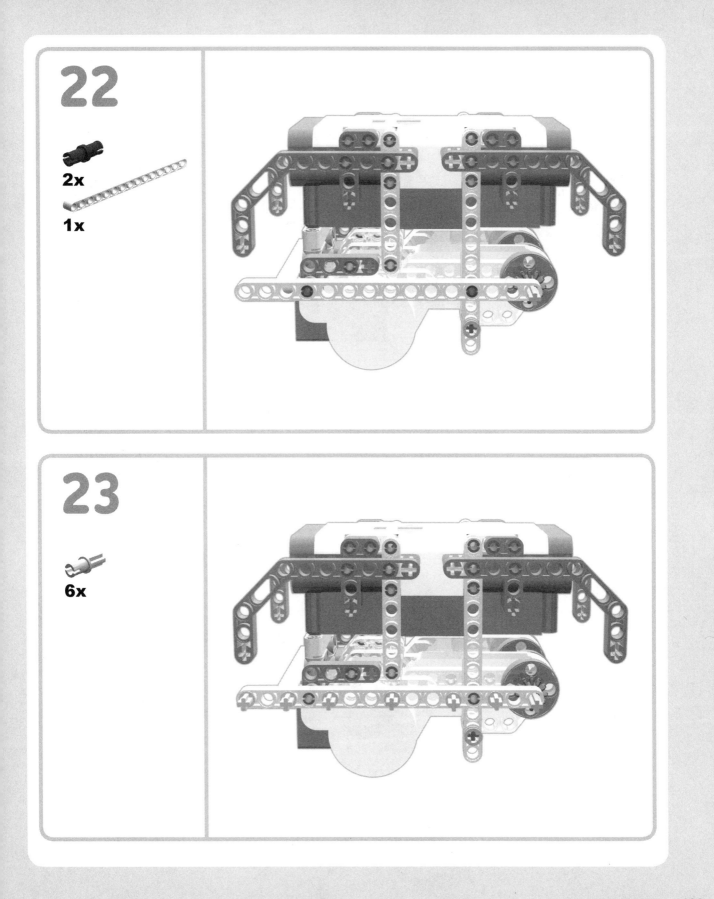

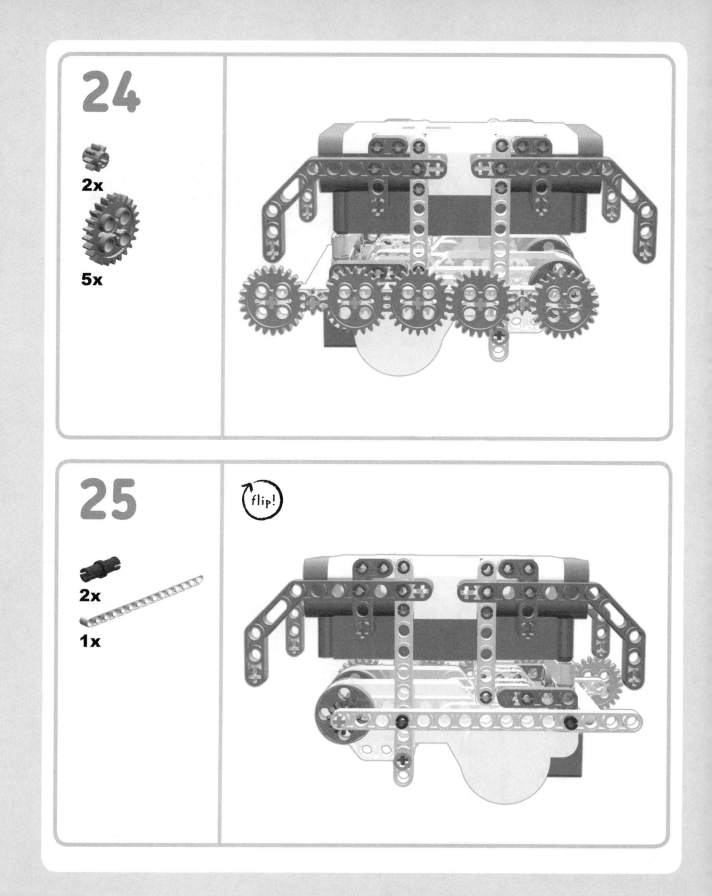

24

2x

5x

25

flip!

2x

1x

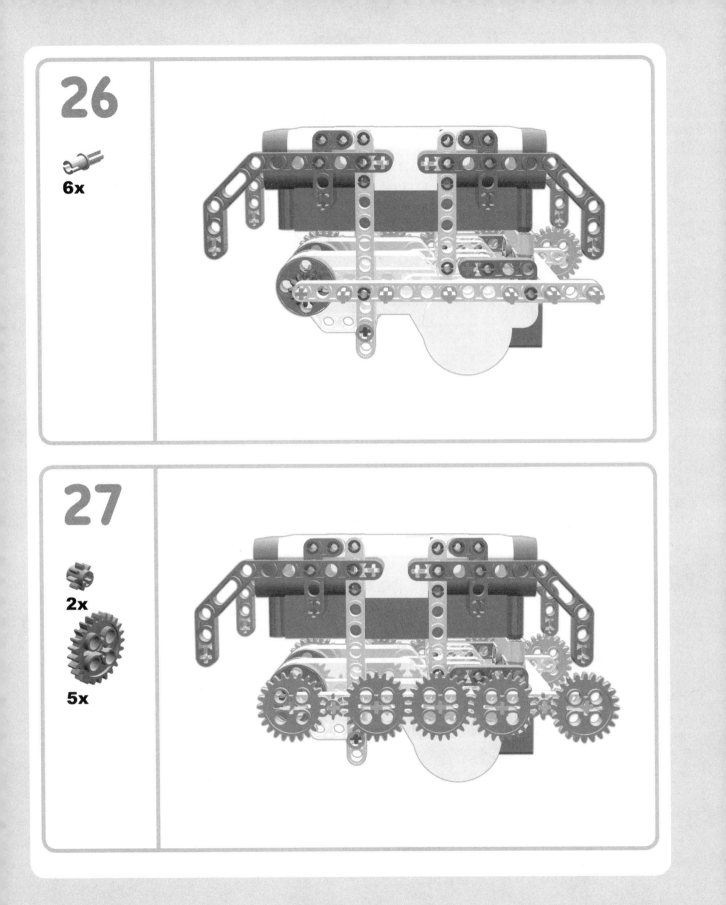

26

6x

27

2x

5x

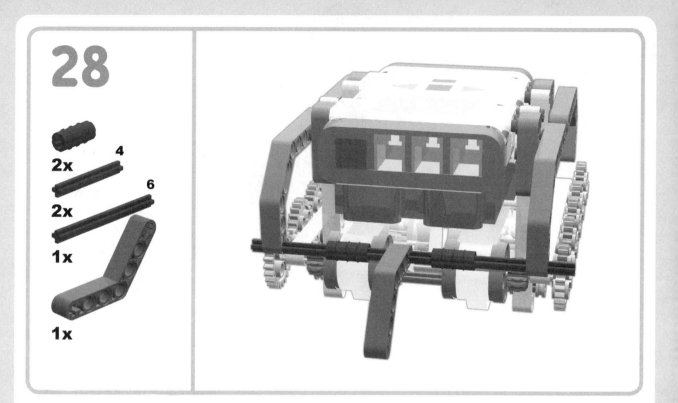

28

2x

4

2x

6

1x

1x

head

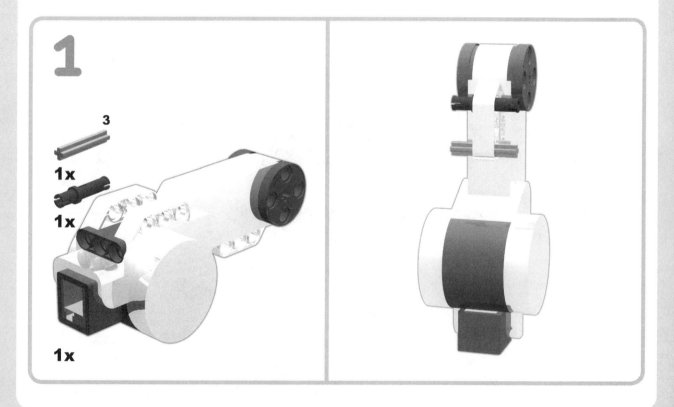

1

3

1x

1x

1x

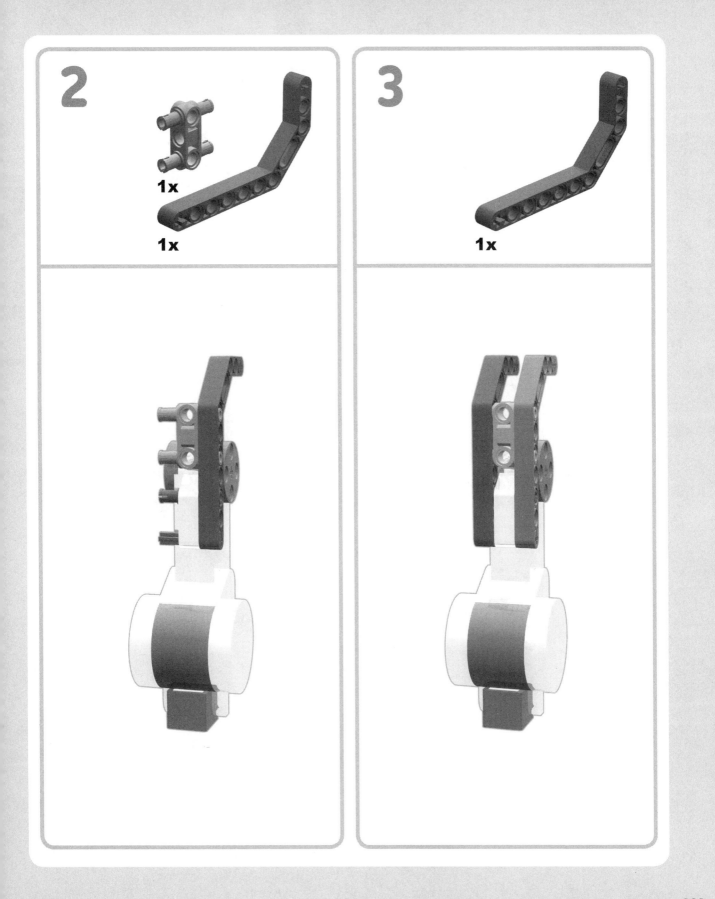

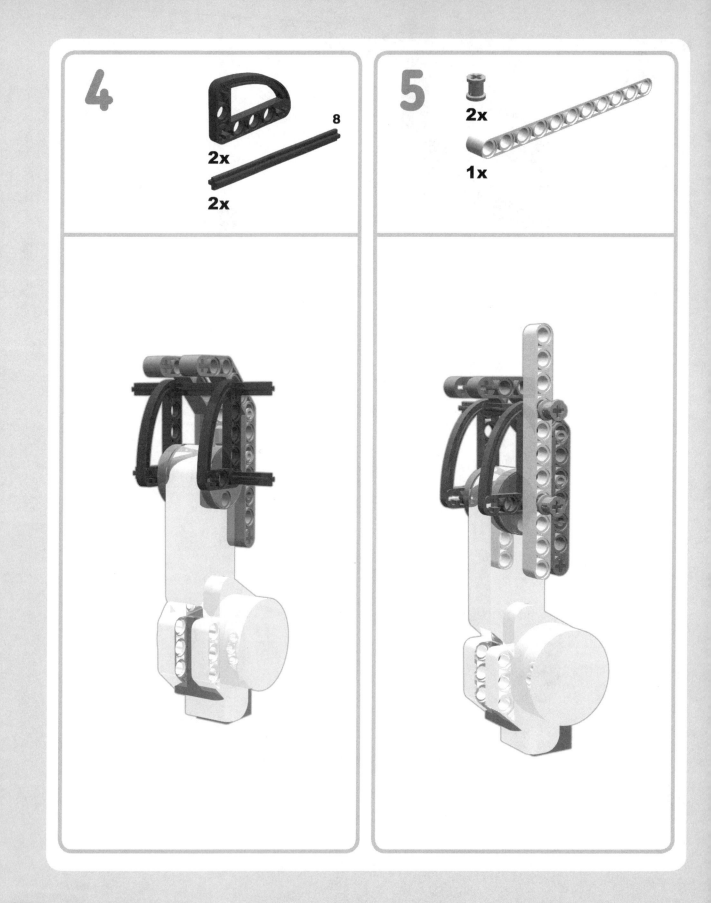

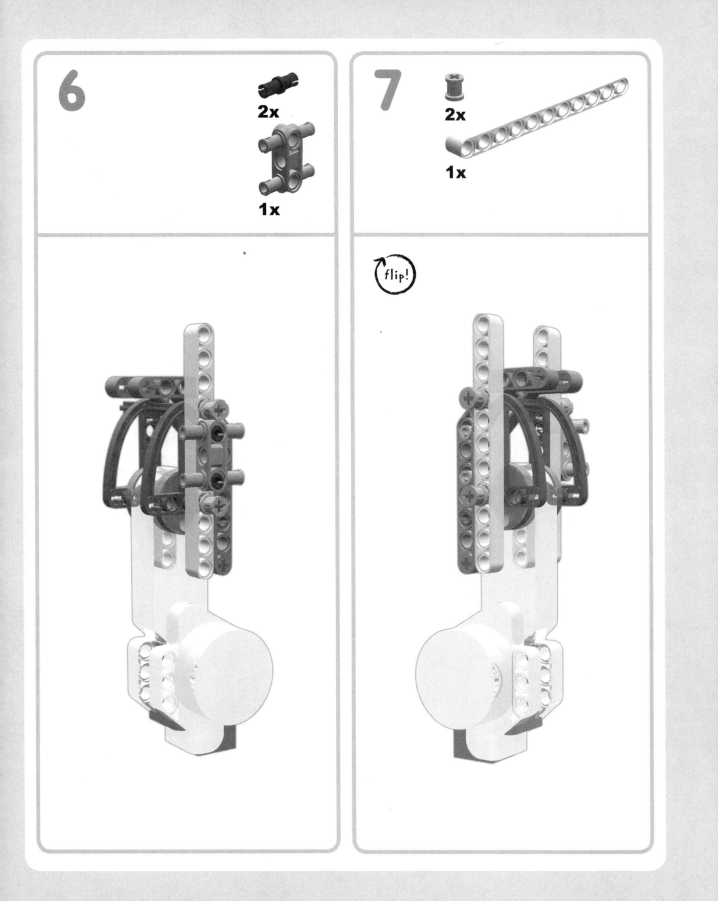

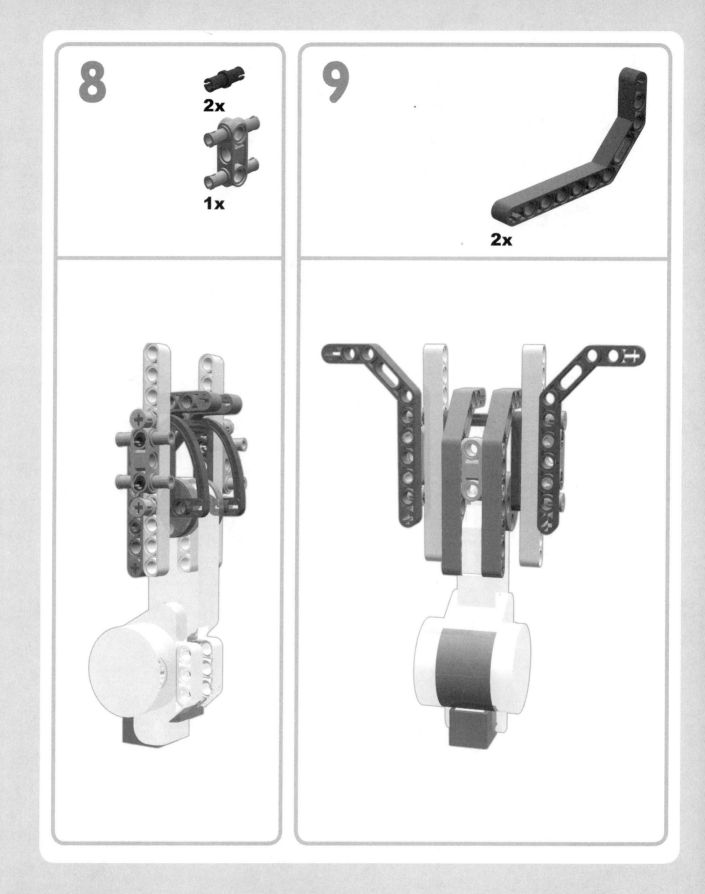

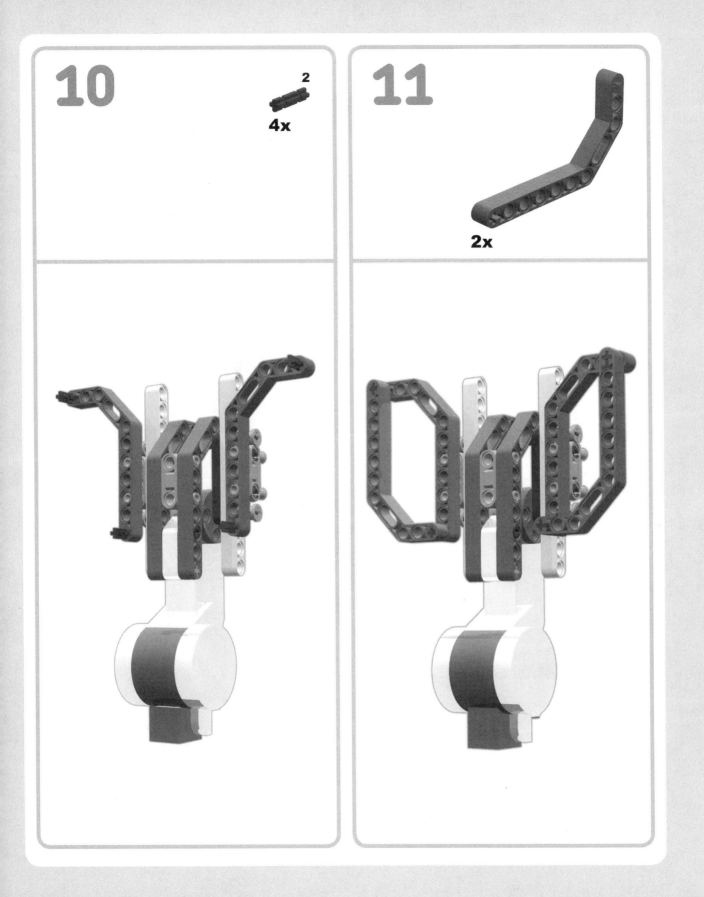

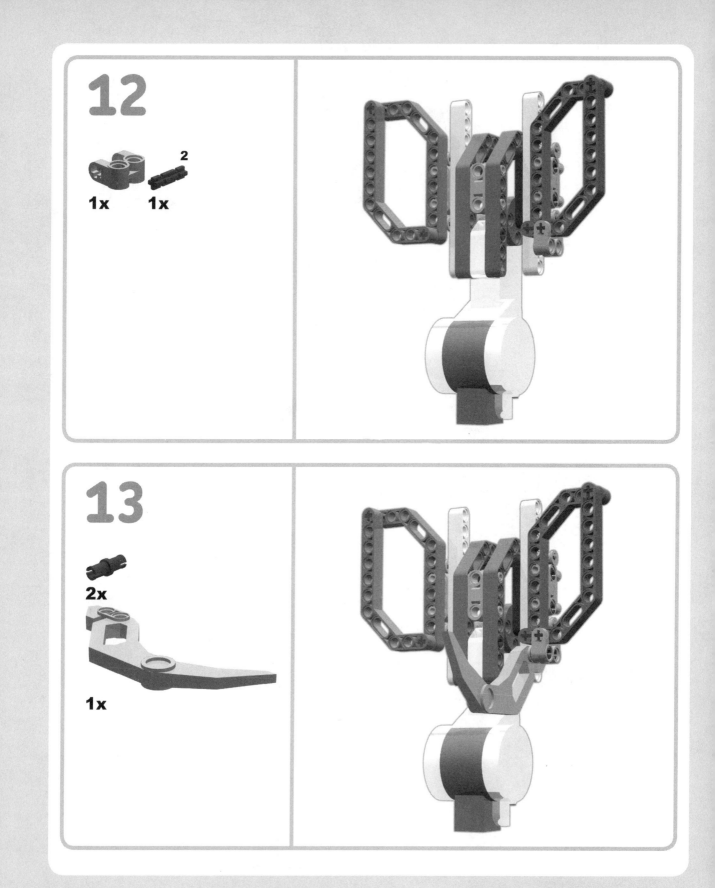

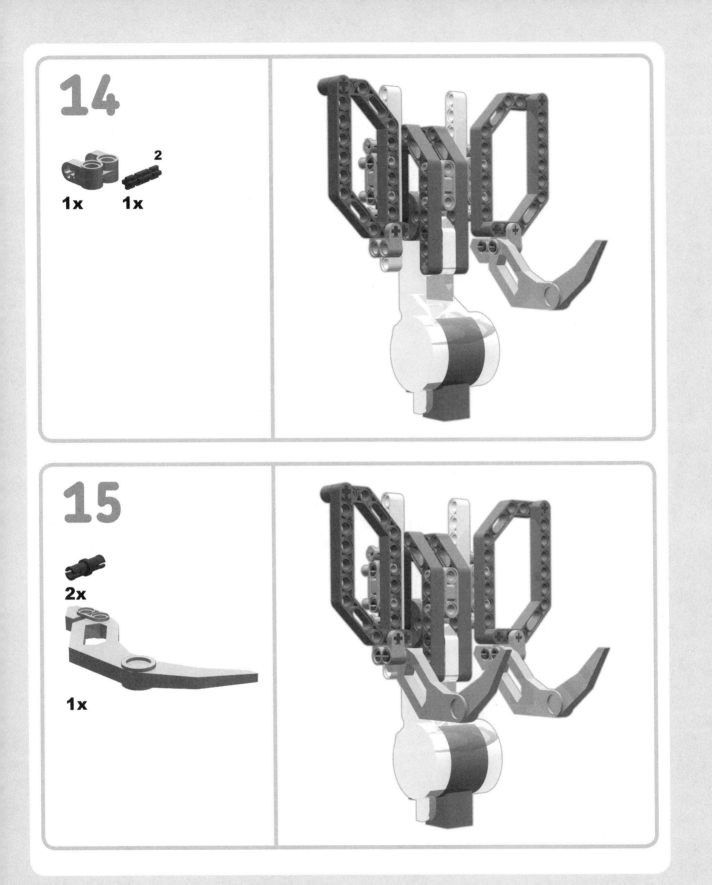

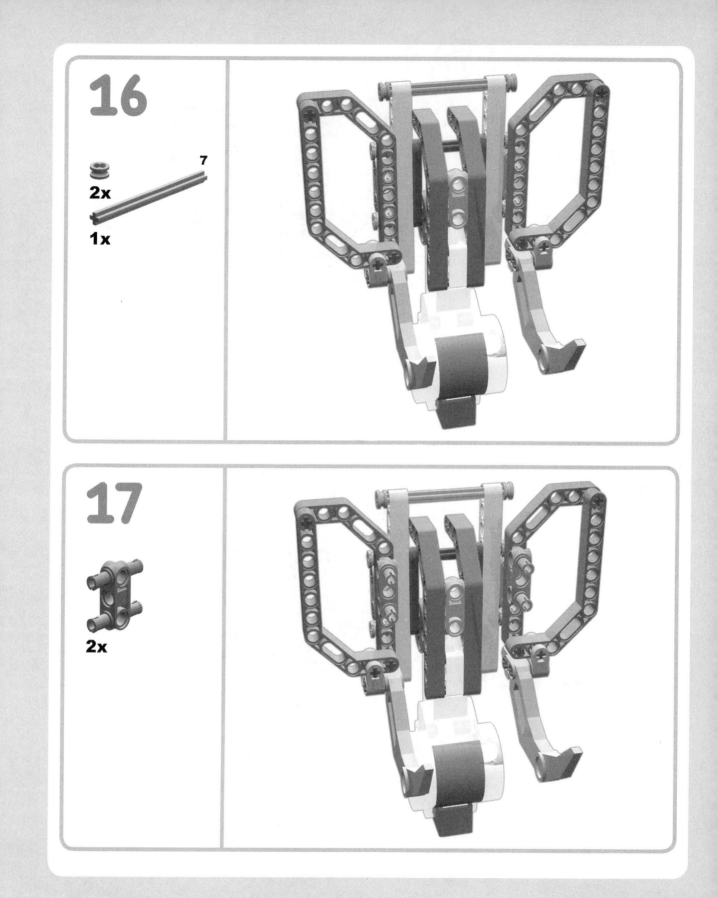

16

2x

7

1x

17

2x

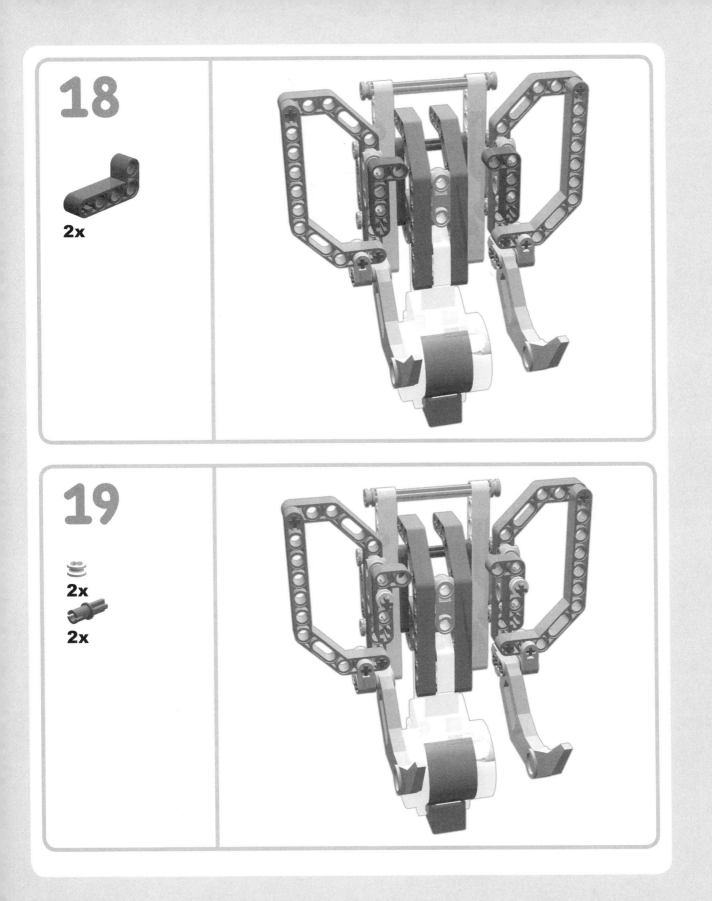

18

2x

19

2x

2x

20

2x

flip!

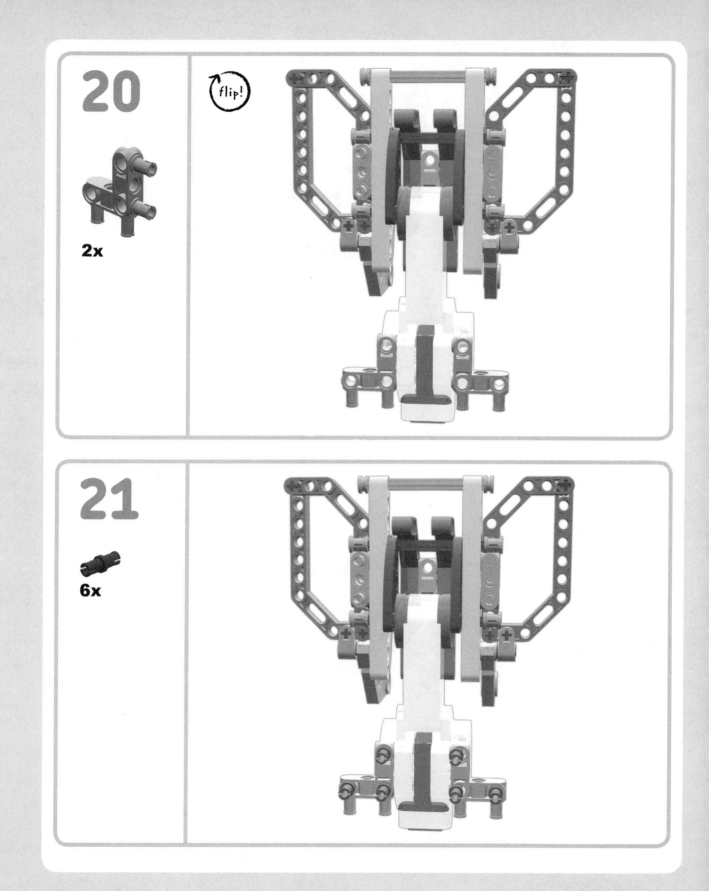

21

6x

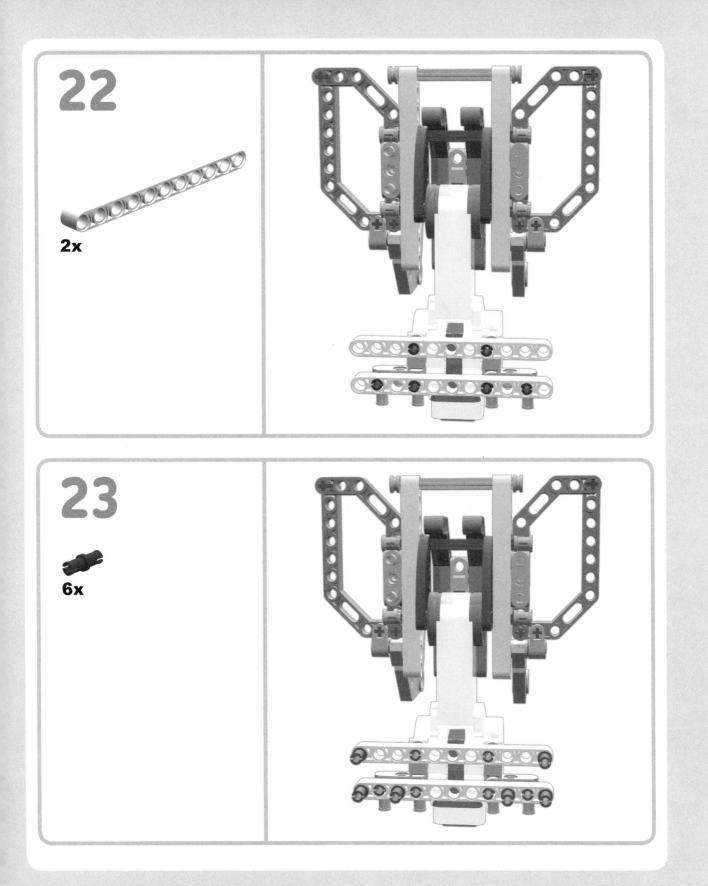

22

2x

23

6x

24

2x

25

4x

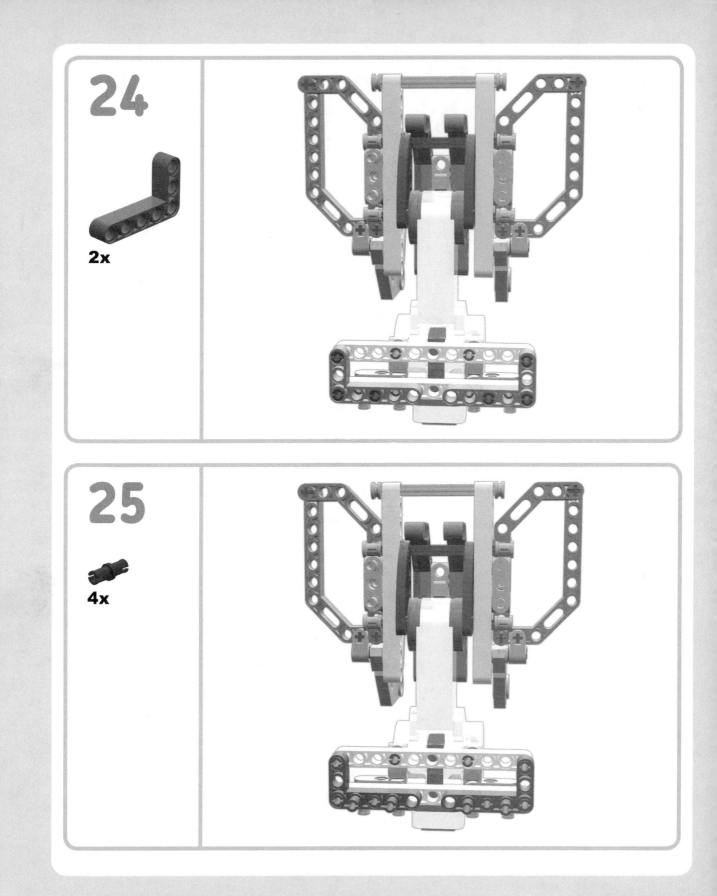

PART SUBSTITUTION SUGGESTION

You can replace the quarter ellipse bent beams with 3 × 5, *L*-shaped bent beams, as shown in Figures 8-2 and 8-3.

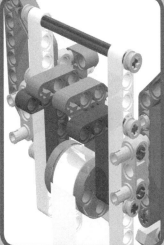

Figure 8-2: In the original elephant head, we used 3 × 5 bent quarter ellipse beams, #8 axles, and regular bushings.

Figure 8-3: In the revised version, we have used the wider 3 × 5 bent beams, #8 axles, and half-bushings.

legs (build four)

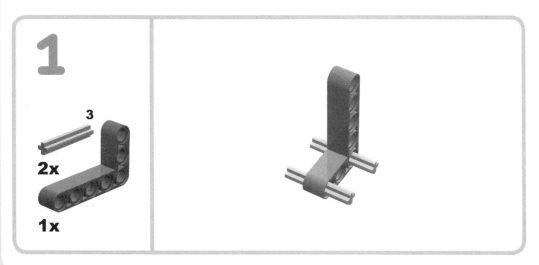

2

2x

3

2x

bottom view

4

1x

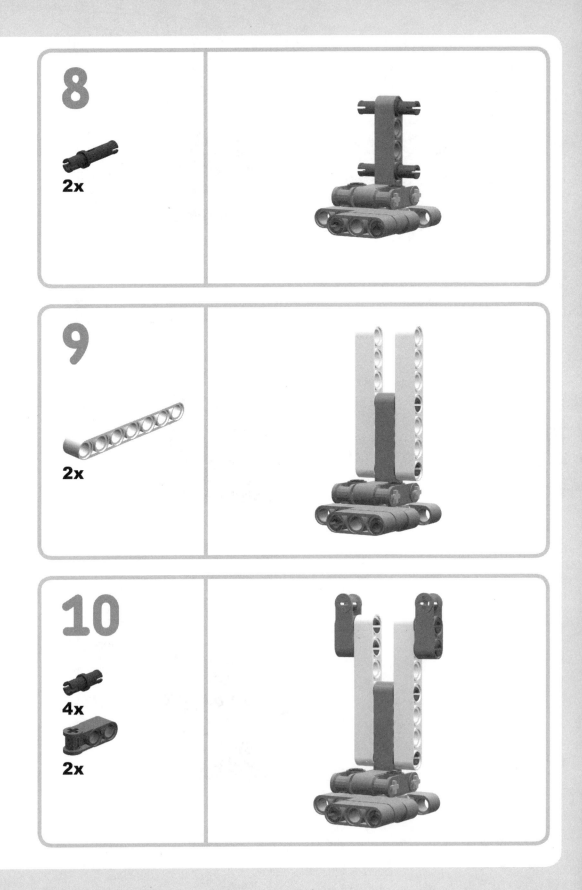

11

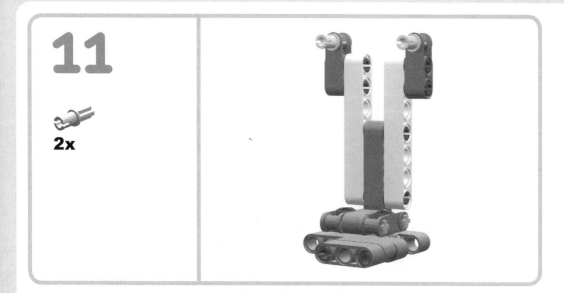

2x

trunk

Cut a long piece of fishing line (at least 24 inches) and thread one end through each of the openings in the piece shaped like a figure eight. Then thread both ends through the seven small cylinders. You will attach this to Pygmy's face.

1x

7x

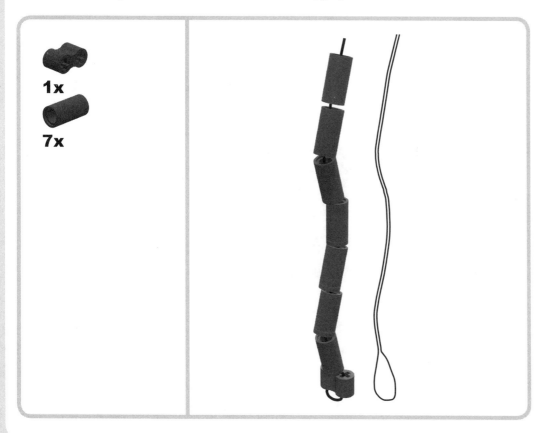

full assembly

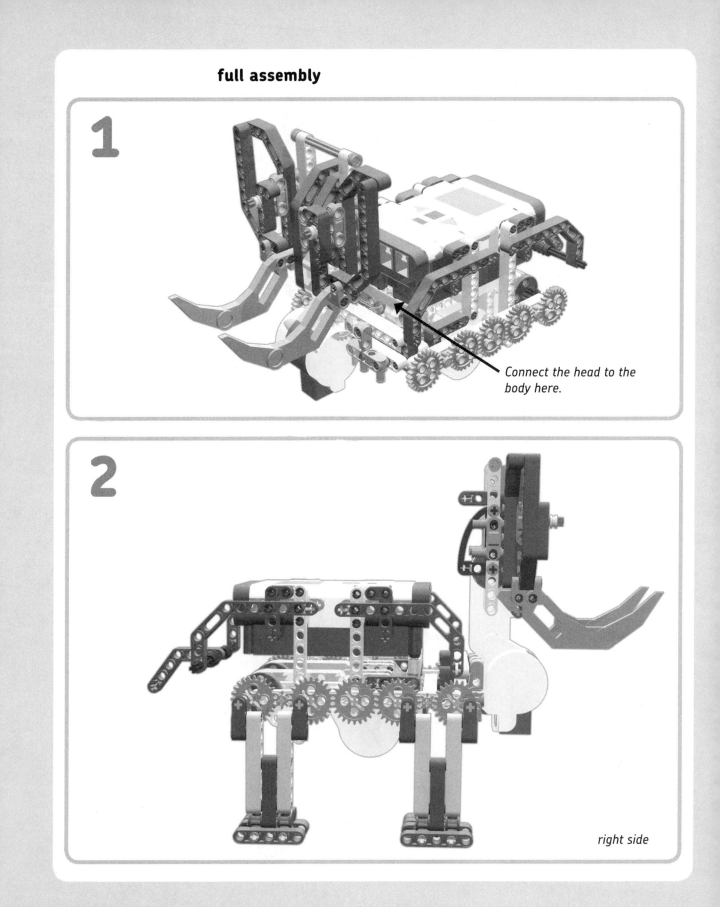

1

Connect the head to the body here.

2

right side

Attach two legs to the right side as shown in step 2. Make sure the gears are lined up exactly as they are here. (The legs are placed in the farthest right and the farthest left holes.)

NOTE You can use the driving gear to adjust the position of the legs on each side.

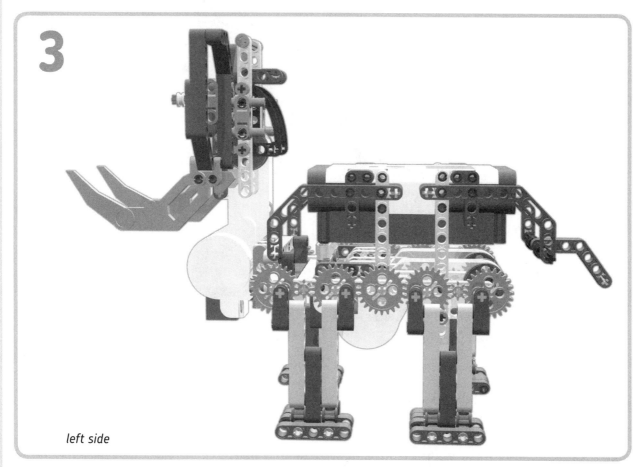

left side

The legs on the left should look like this—both legs should be placed on the inside holes.

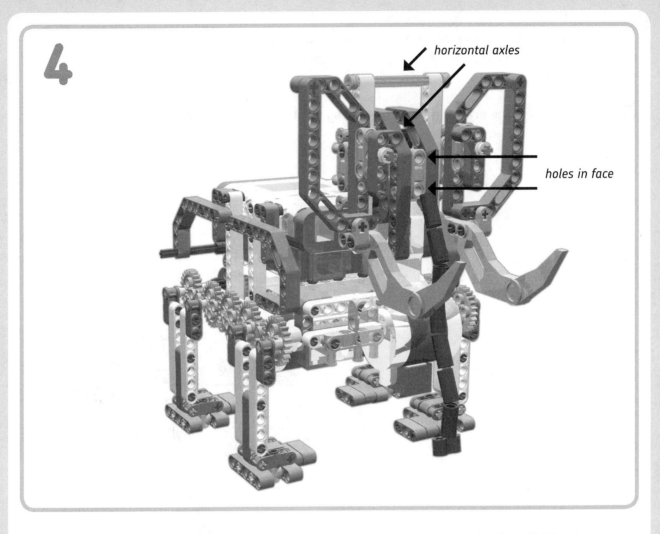

4

horizontal axles

holes in face

Take the trunk you prepared and insert the two ends of the fishing line through one of the two holes in the center of the elephant's face. Then tie it to one of the horizontal axles. You'll need to experiment with how to arrange the trunk and the fishing line. You'll also need to experiment with what is the right amount of tension on the string to get it to curl up when the head is pulled back.

If you use the lower hole, it looks more like the position of the trunk on a real elephant. If you use the upper hole, it seems to lift the trunk better.

wiring connections

Connect the motor in the head to port A. Connect the motors in the legs to ports B and C (it doesn't matter which is which).

programming pygmy

If you've been using the NXT for a while, you are familiar with My Blocks. The purpose of My Blocks is to create miniprograms that can be used repeatedly—and to make long, complicated programs easier to read and understand. In this case, the My Block tells our robot what to do, while the program you will place it in tells the robot when to do it.

my block: elephant

I want Pygmy to walk forward and raise its head and trunk at the same time. This My Block will accomplish that.

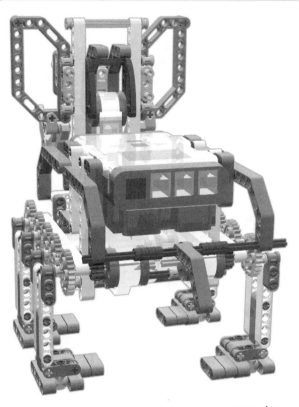

rear view

Figure 8-4: Begin by placing a Move block on the beam. Configure it as shown here.

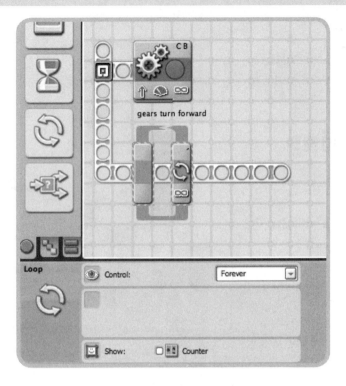

Figure 8-5: Drag down a beam from the orange square at the start and place a forever loop on it.

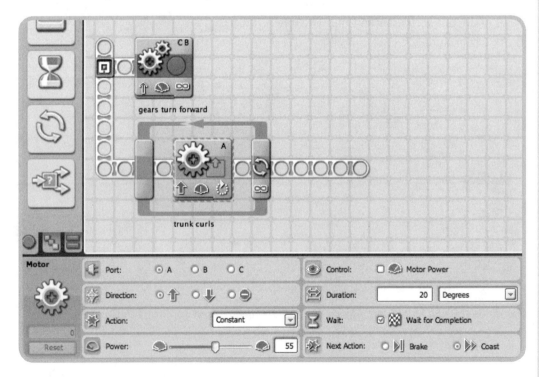

Figure 8-6: Place a Motor block inside the forever loop. Configure it as shown here.

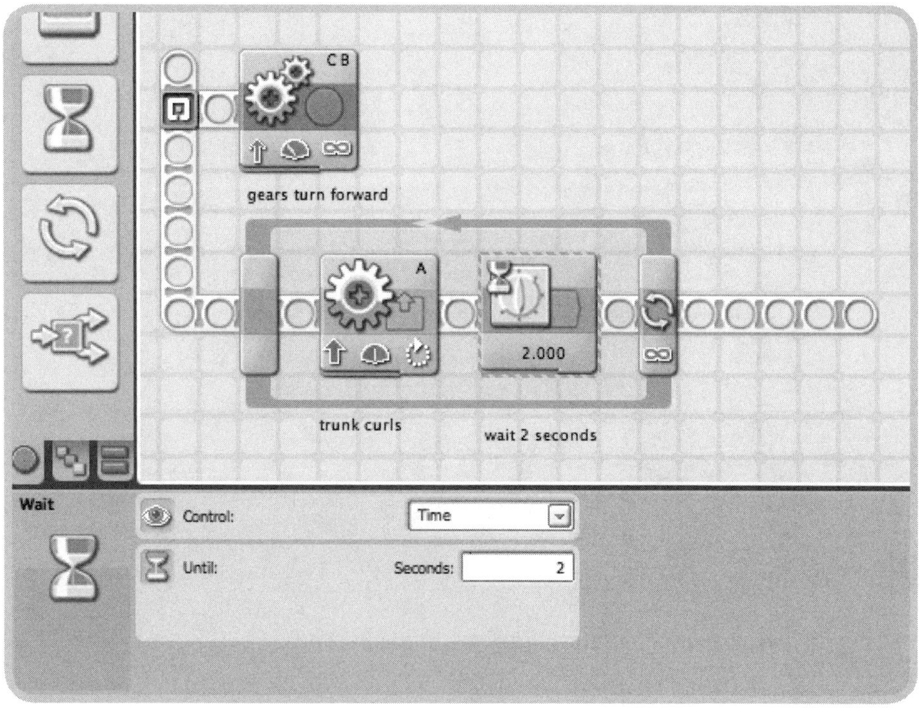

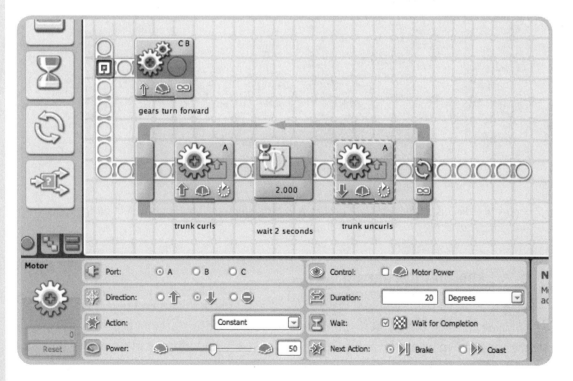

Figure 8-7: Inside the loop, add a Wait block to the right of the Motor block. Set it for 2 seconds.

Figure 8-8: Follow the 2-second Wait block with a Motor block configured as shown here.

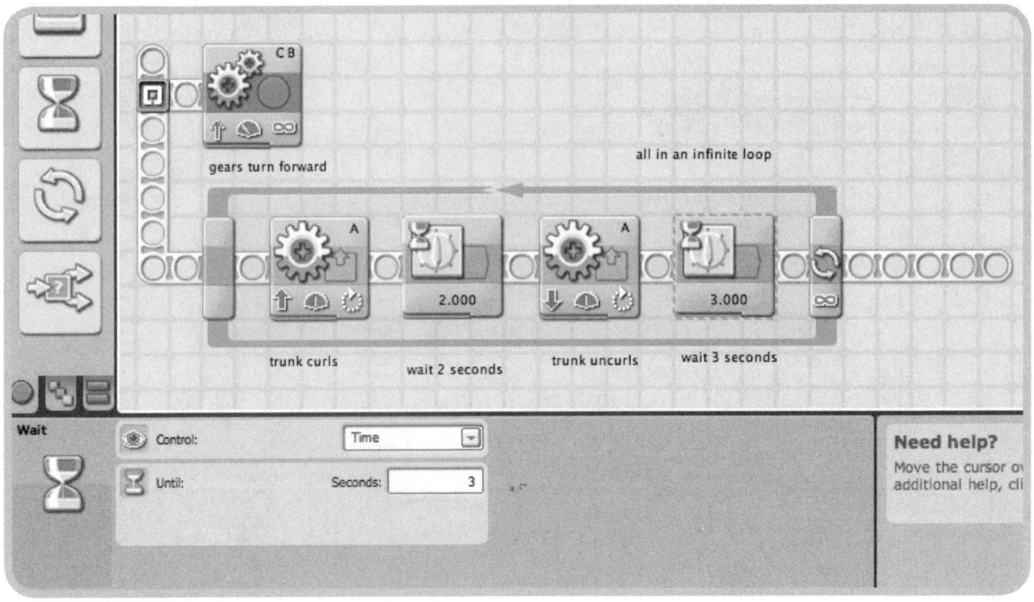

Figure 8-9: Now add a Wait block configured to Time and 3 seconds.

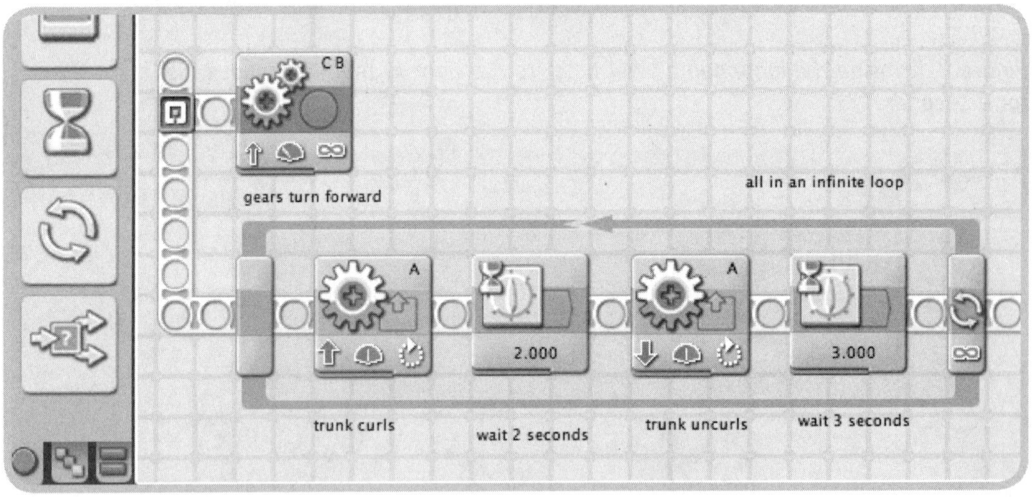

Figure 8-10: This is the complete program for the elephant My Block. Select all the blocks and save it as a My Block called elephant.

the main program

Now that we've told Pygmy to walk forward and raise its head and trunk, we want to tell it when to do this. First, I want to give myself time to start the program and set the robot down where I want it. The first block sets that amount of time.

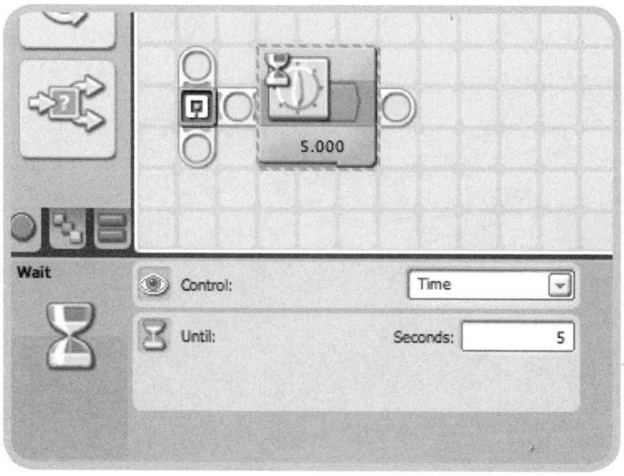

Figure 8-11: In a new window, place a Wait block on the beam and set it for 5 seconds.

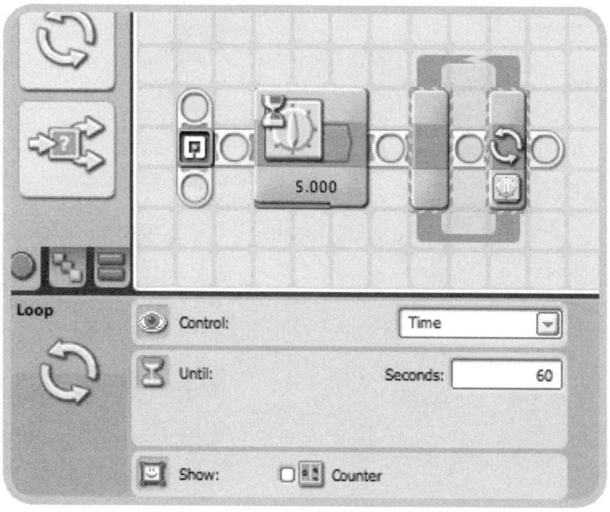

Figure 8-12: Next, place a loop on the beam. Configure it to Time and 60 seconds.

You'll notice that I have not added sound effects to Pygmy, but the sound of a trumpeting elephant would be an excellent addition. It's not hard to do, and iTunes offers a number of different options. You will find directions for creating custom sound files in the MINDSTORMS section of the LEGO website. (Search for the words *custom sounds* at *http://mindstorms.lego .com/nxtlog/default.aspx*.) There are also directions on this book's companion website (*http://www .thenxtzoo.com/*).

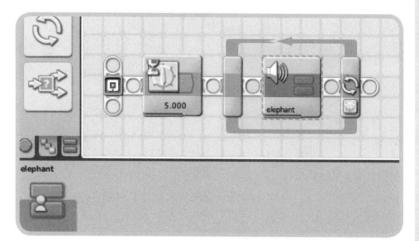

Figure 8-13: Finally, go to the My Block palette, select the elephant My Block you just created, and place it in the loop.

This robot runs best if the power is kept rather low, so as you test the completed elephant, be aware that you will need to adjust the power settings.

Also, if the head is not raising back and pausing as commanded, try tying the trunk to different places on the head. There is no single perfect place, but you might find a spot where you prefer the way the trunk curls.

polecat: an NXT skunk

Watch out for this critter! If you get in its way, it'll have a surprise for you.

Figure 9-1: The NXT skunk

My first choice was for Polecat to spray perfume, but perfume is a liquid—and that is not compatible with electronics. Instead, I have used the competition cannon and arrow as a method of attack.

If you've seen videos of Polecat in action, you might have seen it with the optional fur on its tail. This is artificial fur from a craft store. While it looks very cool, the additional weight does affect the turning accuracy of the robot. The driving wheels are in front, so any weight you add to the back reduces traction for the robot in general.

You might consider programming your Polecat to shoot its dart randomly, so no one knows when it will happen. Whenever it happens, it always provokes a laugh—unless, of course, you shoot it under the stove, like I did once. (Boy, that was hard to get out!)

building polecat

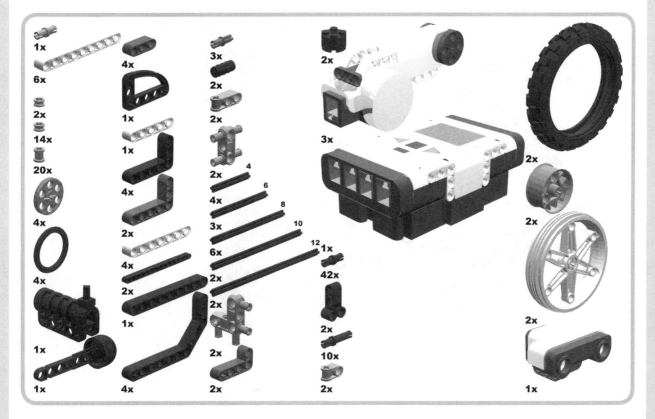

* Non-LEGO part required: Fishing line
* Optional: Artificial fur, available at all major craft stores

body

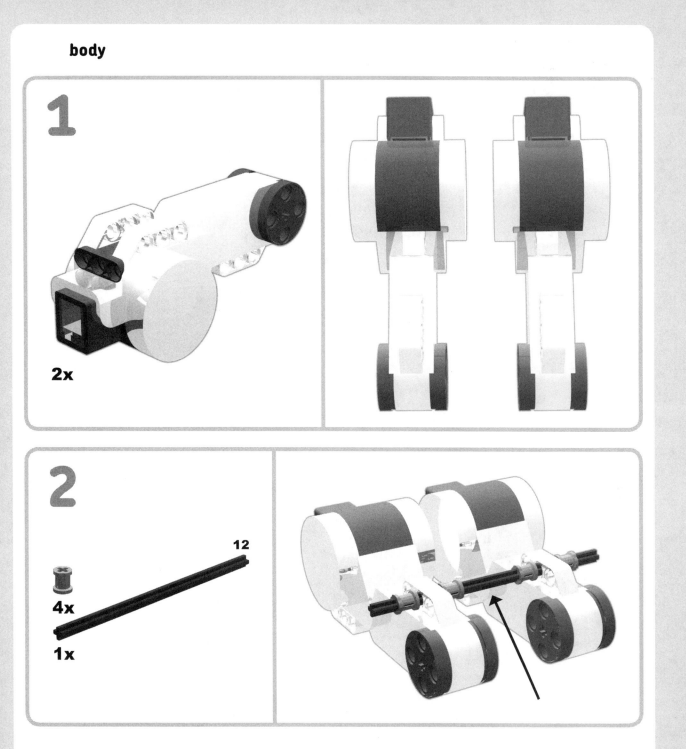

The space between these bushings is one inch.

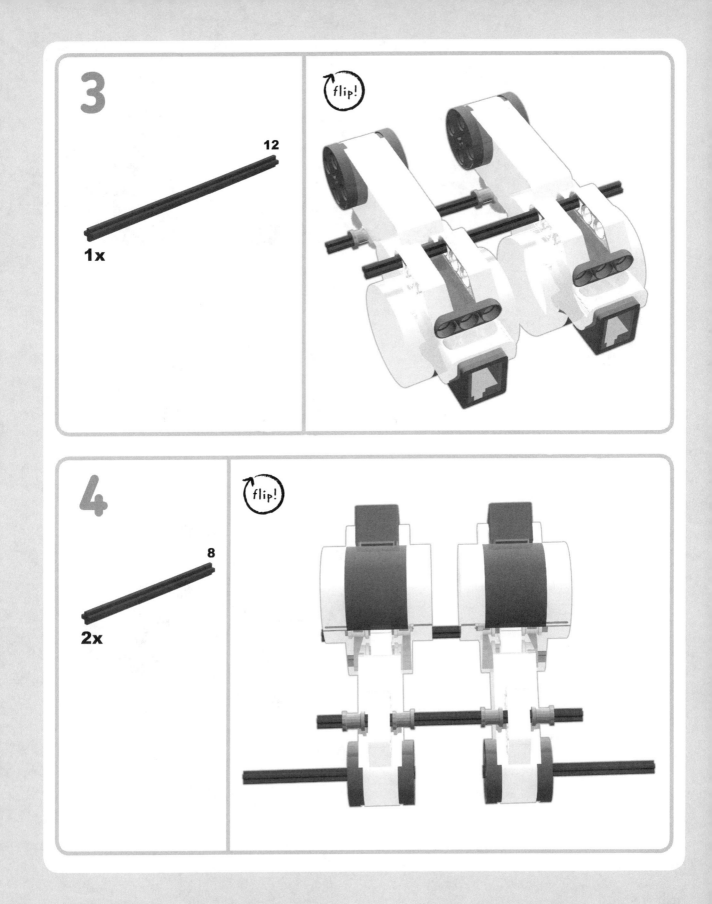

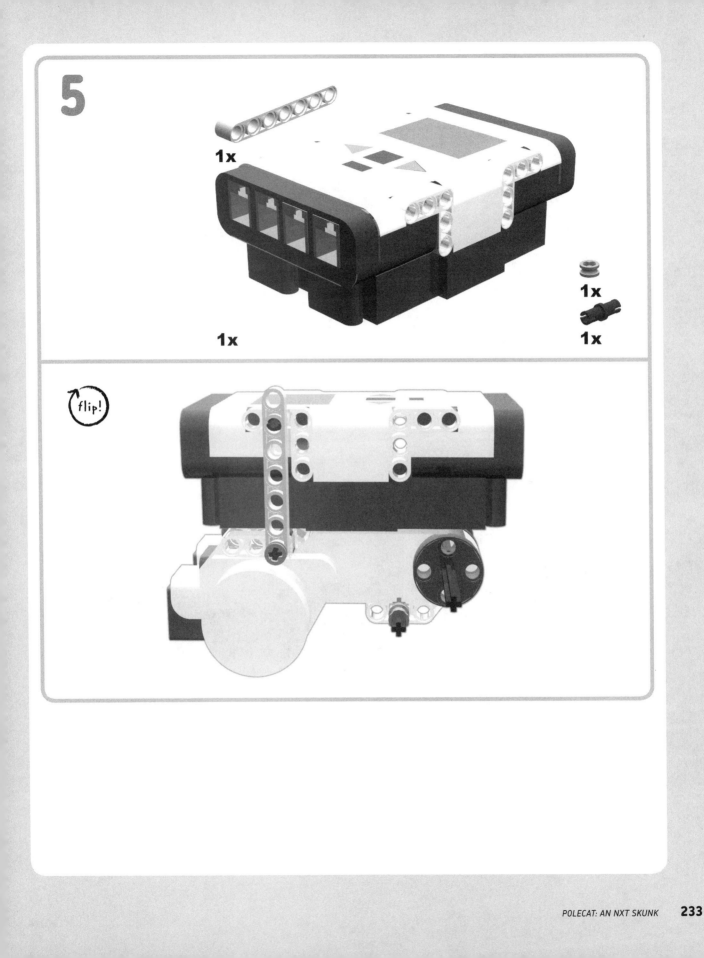

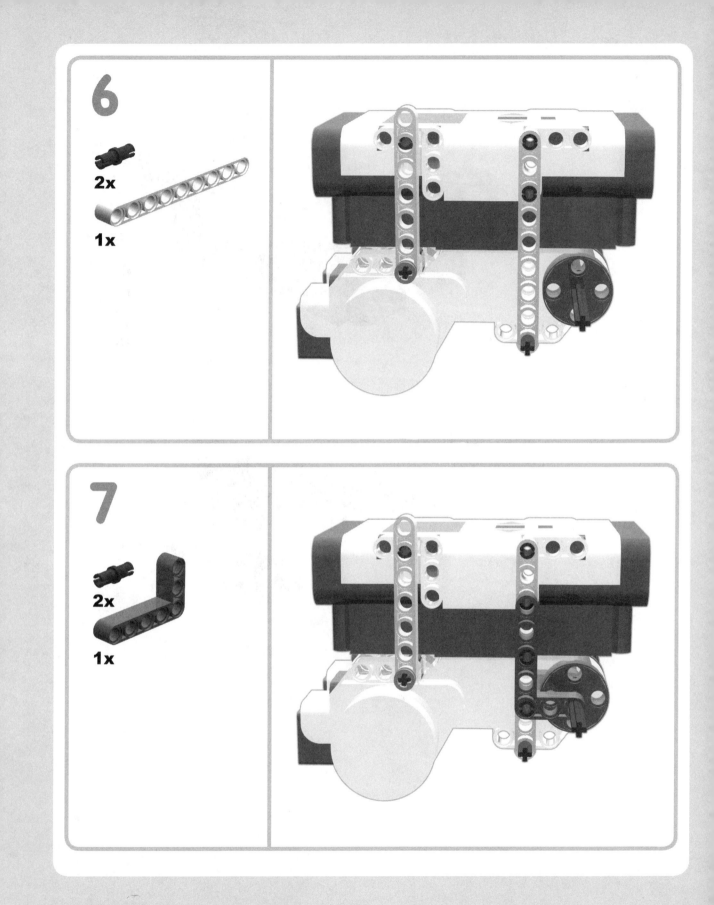

8

2x

1x

flip!

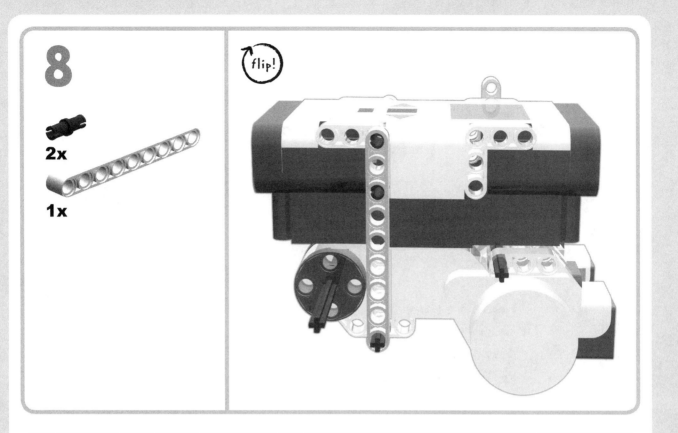

9

1x

1x

1x

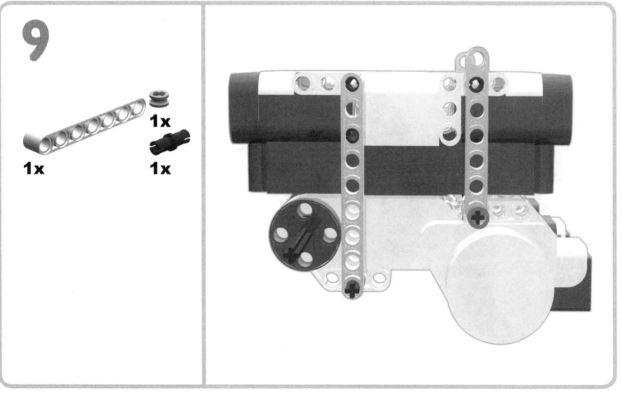

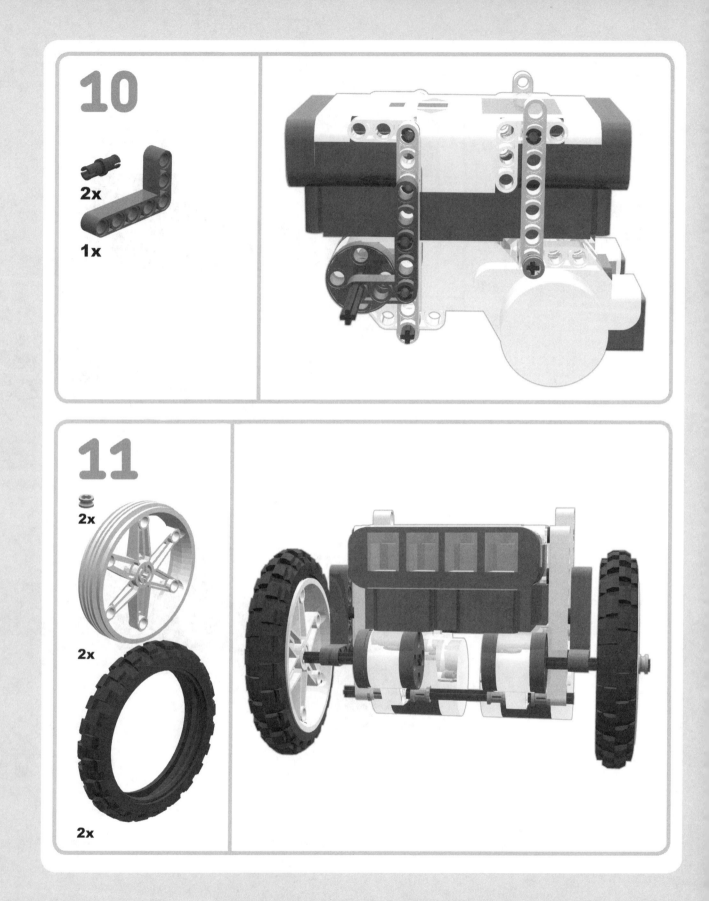

head

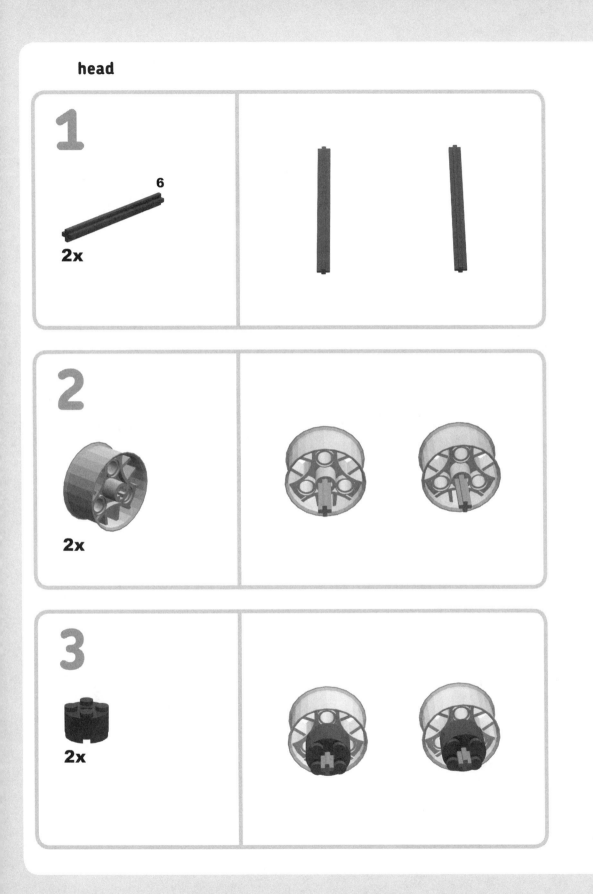

1

2x

6

2

2x

3

2x

4

2x

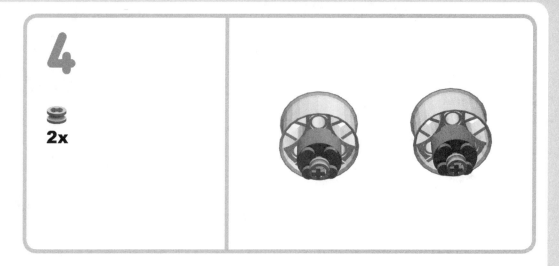

5

1x

flip!

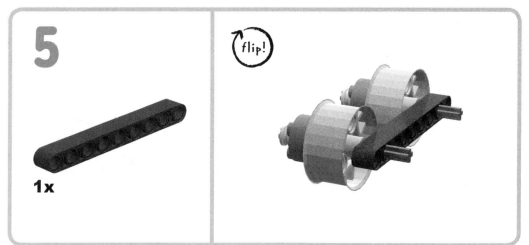

6

2x

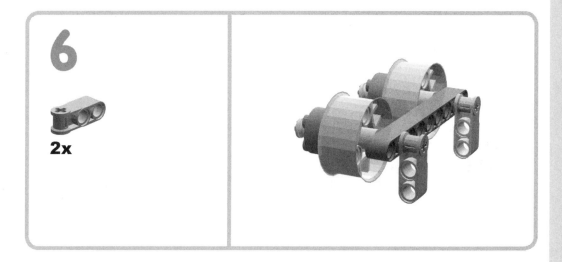

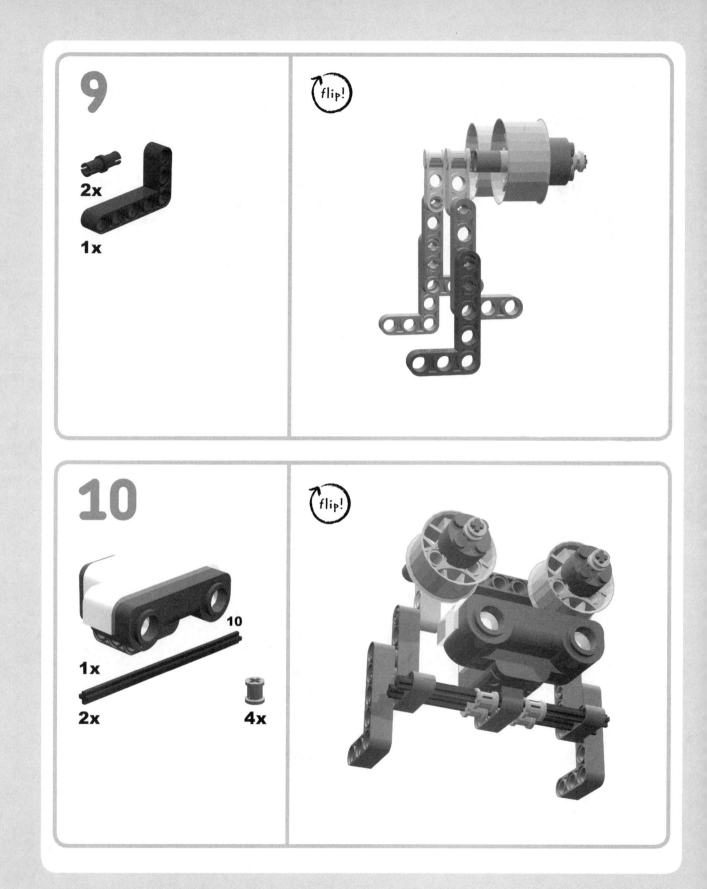

11

4x

tail

1

1x

1x

1x

1x

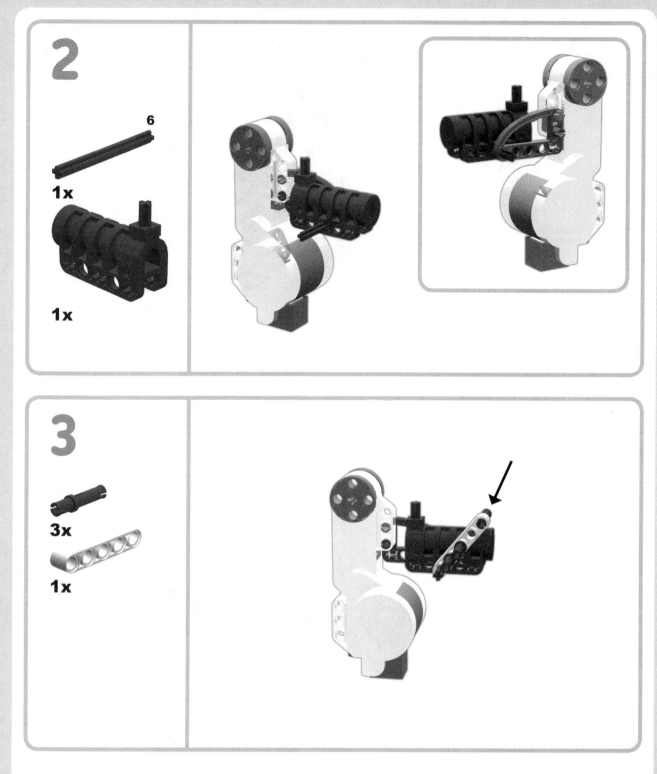

This pin faces inward. It will pull the trigger.

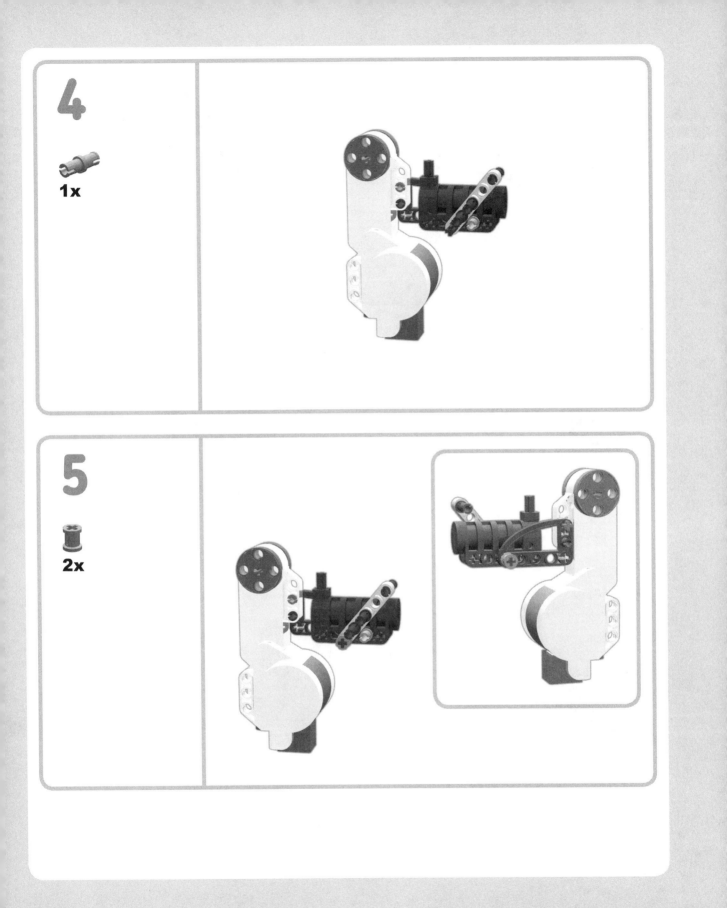

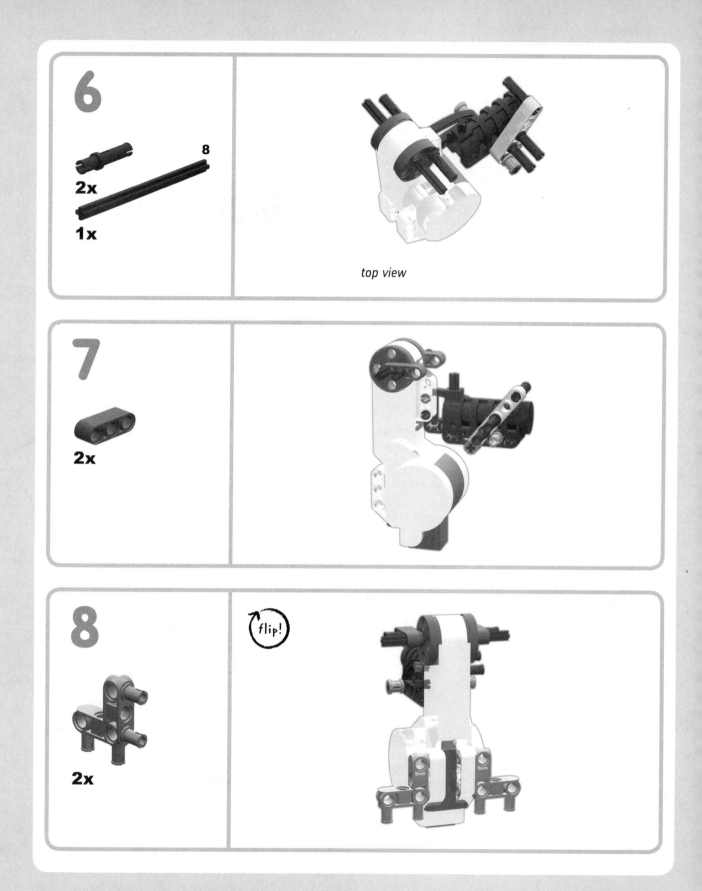

top view

flip!

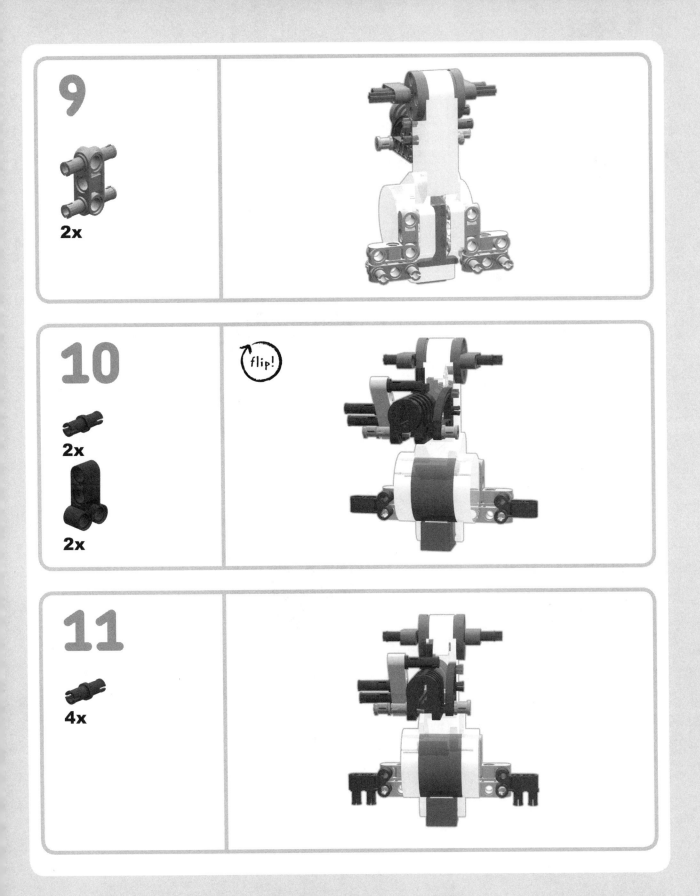

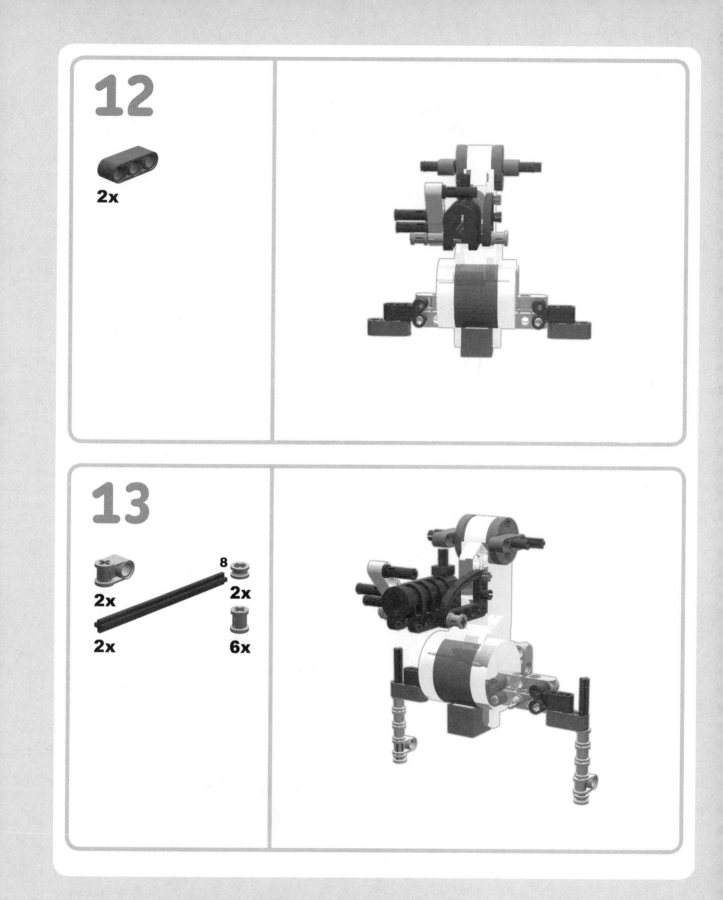

14

2x

15

4x **4x** **4x**

4x **4**

2x

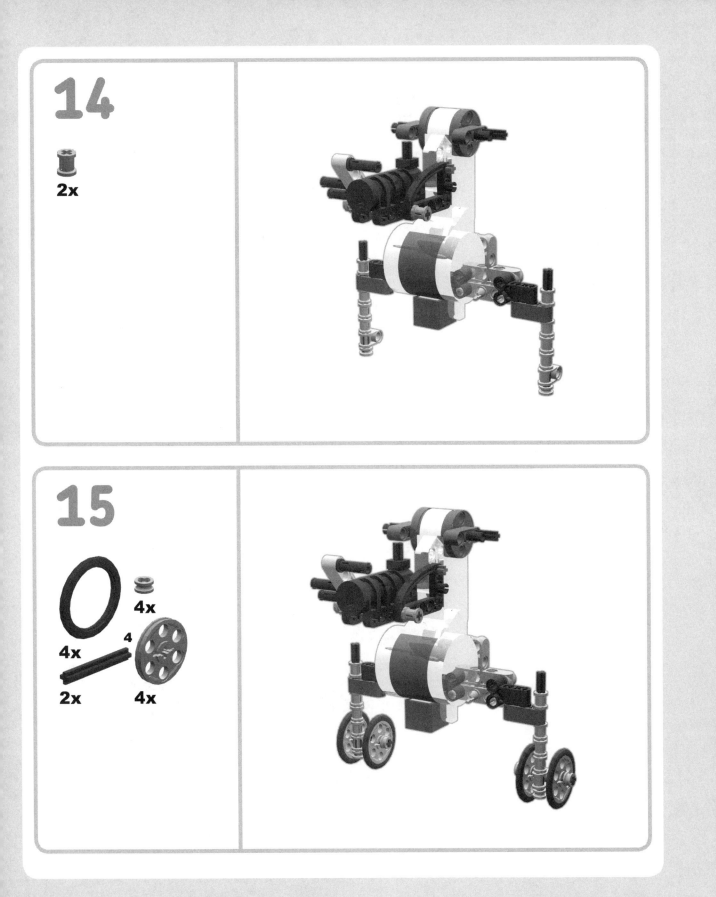

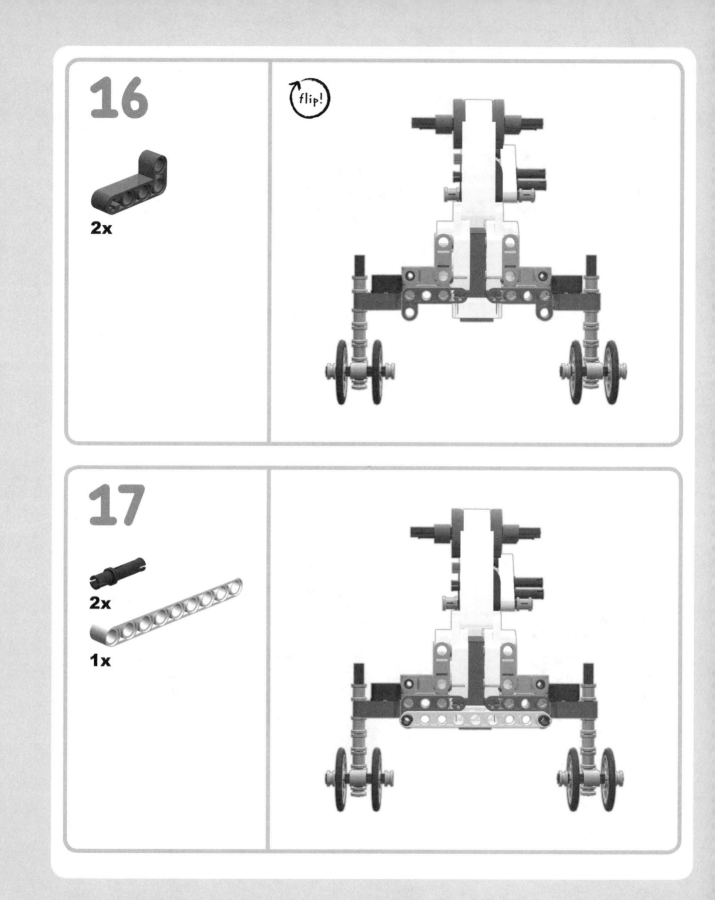

16

2x

flip!

17

2x

1x

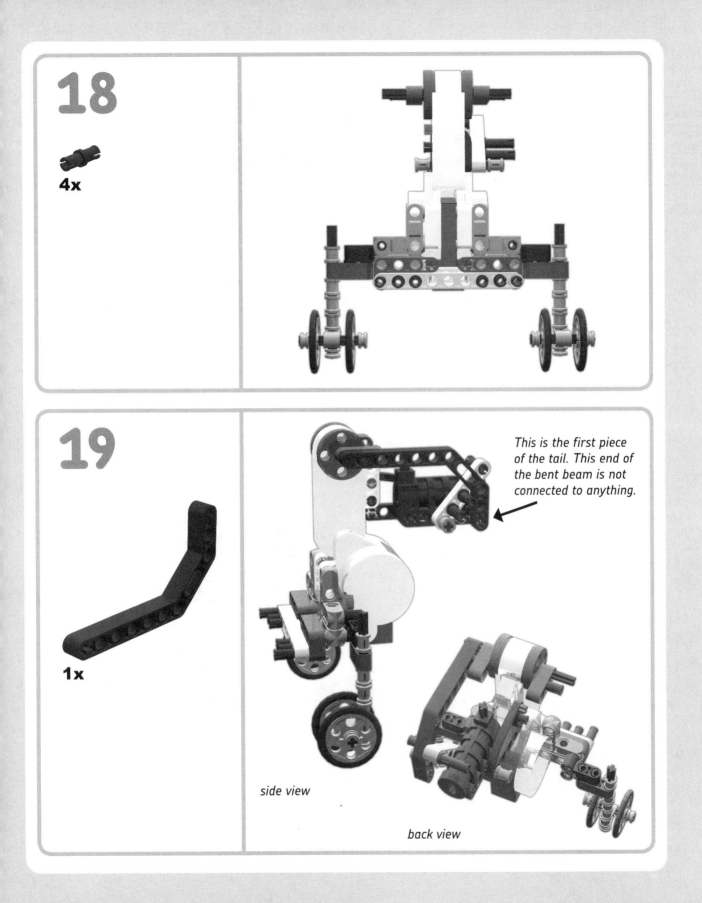

18

4x

19

1x

This is the first piece of the tail. This end of the bent beam is not connected to anything.

side view

back view

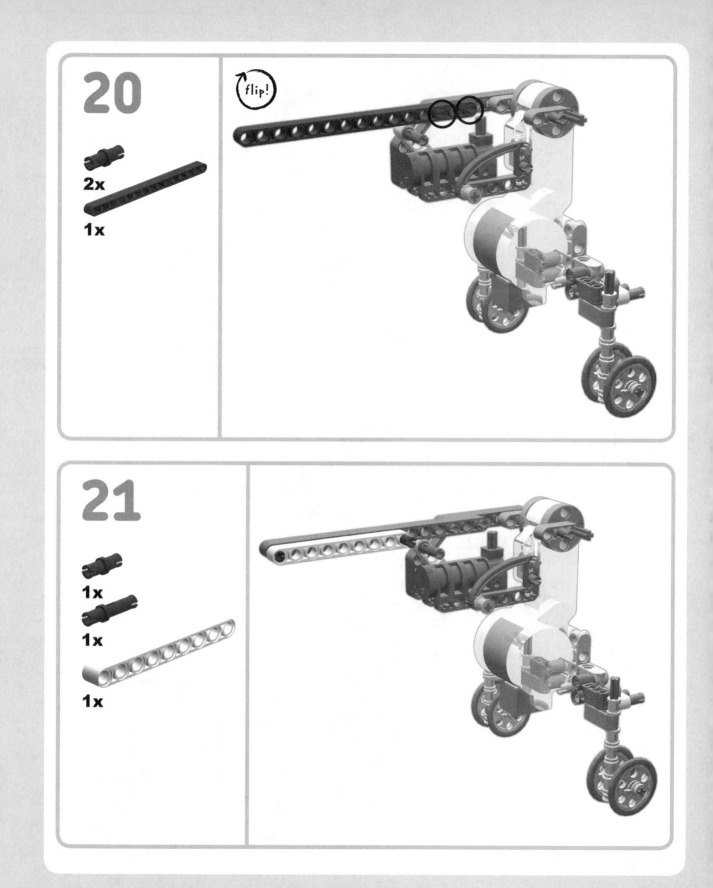

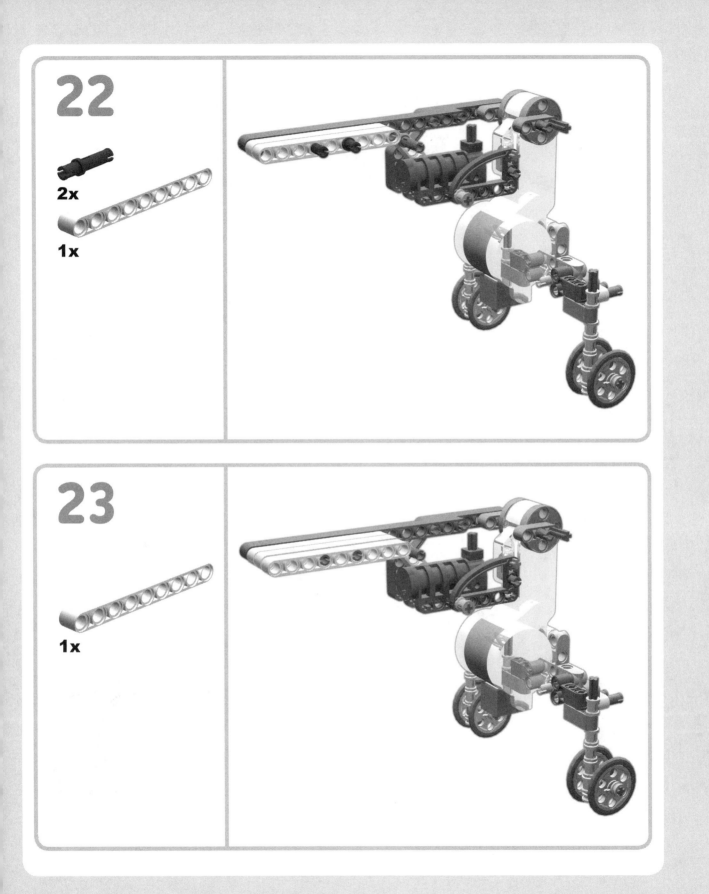

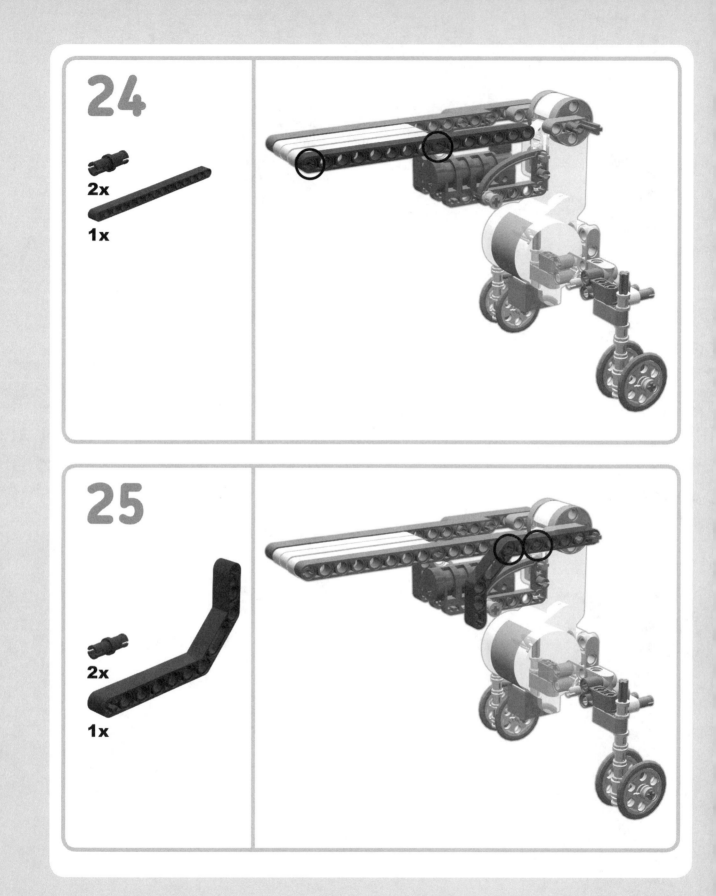

24

2x

1x

25

2x

1x

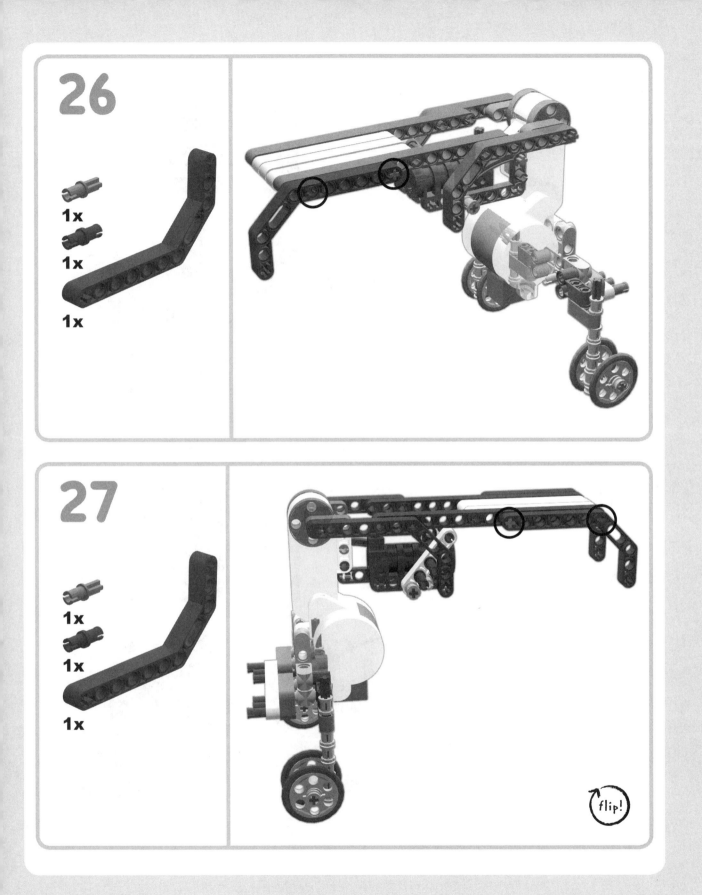

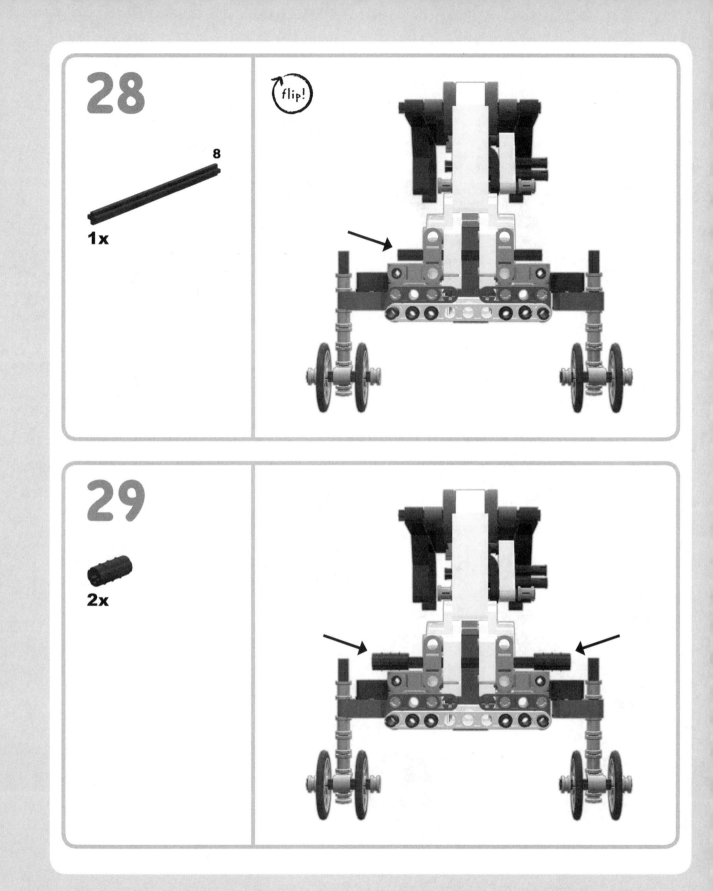

30

2x

4

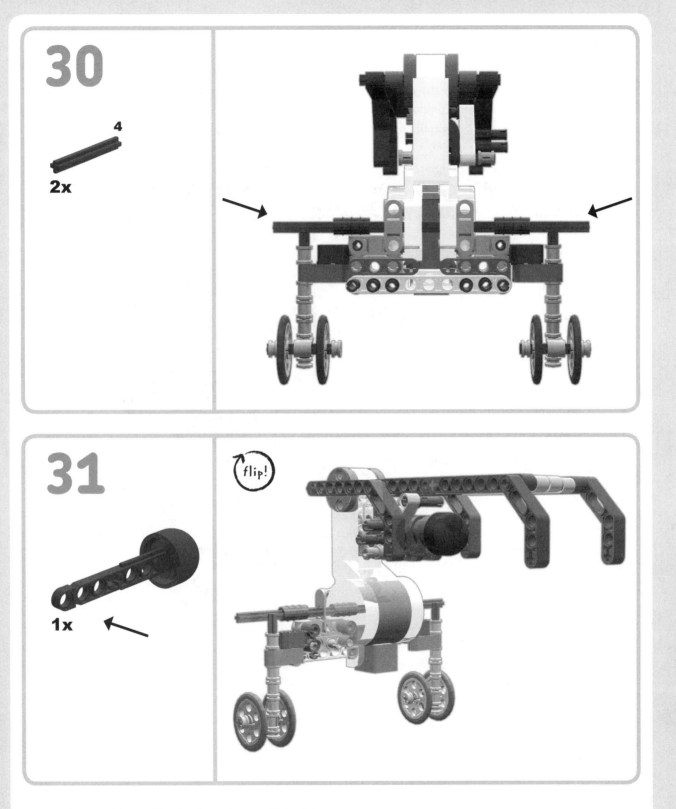

31

flip!

1x

Insert this end in the launcher until it clicks.

PART SUBSTITUTION SUGGESTION

The skunk can be built without the cannon and its missile, but that would be a shame. It never fails to surprise and bring a laugh to observers.

This robot uses a lot of half-bushings, but you can experiment with using regular (wider) bushings.

As with Pygmy the elephant, the 3 × 5, L-shaped beams in the tail can be substituted for the 3 × 5, L-shaped beam with a quarter ellipse, as shown in Figures 9-2 and 9-3.

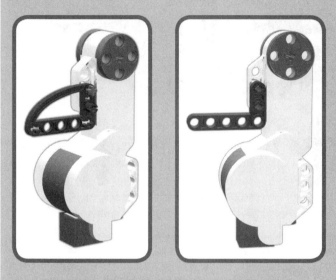

Figure 9-2: This 3 × 5, L-shaped quarter ellipse (shown on the left) can be replaced with the simple 3 × 5, L-shaped bent beam. If you make this change, you will also need to modify the connectors used with it, because the replacement beam does not have axle holes like the original beam.

Figure 9-3: With the substitute beam, you will need to hold the cannon on with a long black pin, a #4 axle, and a bushing. (This beam does not have axle holes like the original quarter ellipse beam.)

full assembly

1

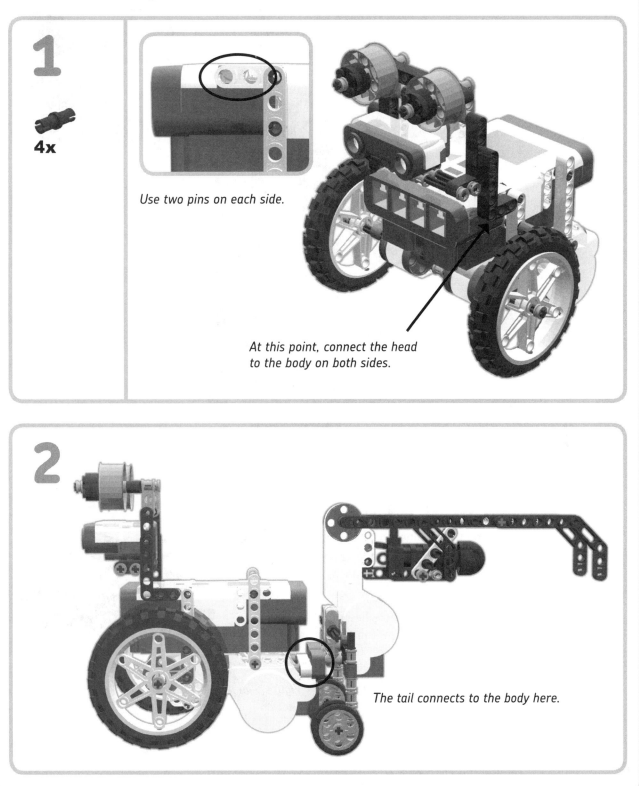

4x

Use two pins on each side.

At this point, connect the head to the body on both sides.

2

The tail connects to the body here.

2x

2x   **2x**

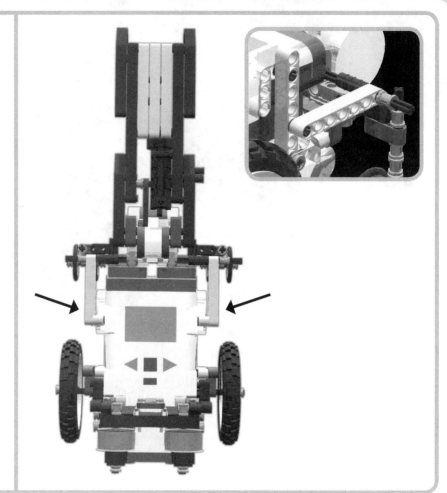

connecting the cannon trigger

Using the fishing line, connect the beam hole to the pin, as shown in Figure 9-4. (There is a slit in the pin you can tie the line to.) When the tail raises, the 5-hole beam should pull backward and press the trigger.

wiring connections

Connect the Ultrasonic Sensor to port 4. Connect the two wheel motors to ports B and C. Connect the tail motor to port A.

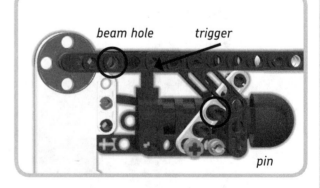

beam hole   *trigger*

pin

Figure 9-4: A detailed view of the cannon trigger

programming polecat

We broke the programming for Polecat down into three subsections. First, we'll create a My Block to manage the tail action. Next, the main program contains the complete directions for what you want Polecat to do. Finally, the third program allows you to tell Polecat when you want it to act—in particular, how many times or how long you want it to repeat the main program.

my block: tail

This My Block will control the motion of the tail—raising and lowering it.

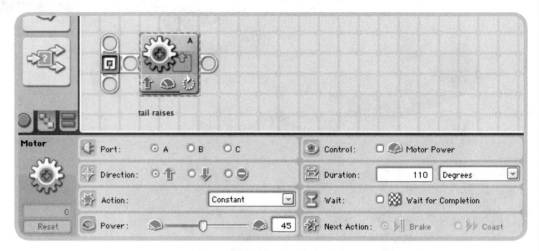

Figure 9-5: To raise the tail, we will use a Motor block. Configure the block as shown in this figure.

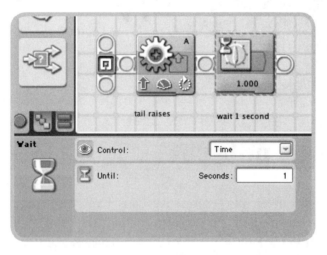

Figure 9-6: We'll then use a Wait block to tell the tail to pause for 1 second. Configure it as shown.

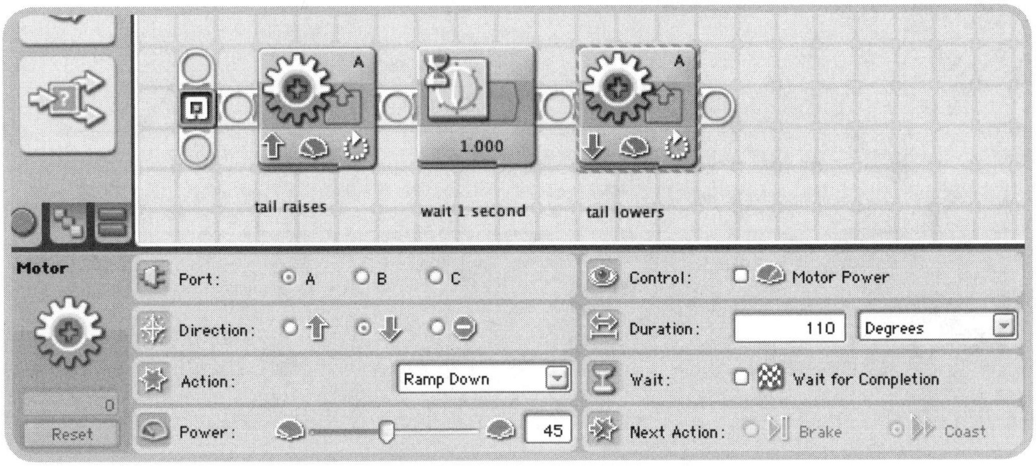

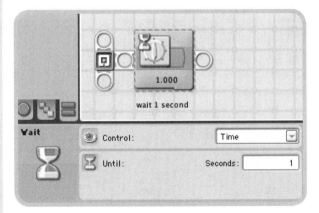

Figure 9-7: Add a Motor block configured as shown here, and your My Block will be complete. Select all the blocks and create a My Block named tail.

the main program

This My Block contains all the actions of Polecat.

Figure 9-8: The first block in this program is a Wait block configured to 1 second.

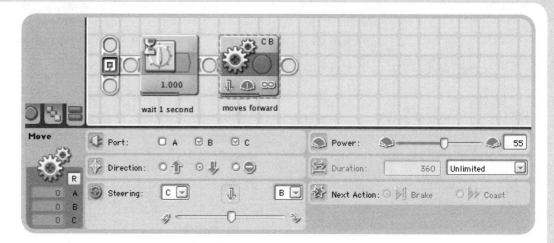

Figure 9-9: Next, we will add a Move block configured for motors B and C.

Sometimes you need to choose the backward arrow to make a robot move forward. This is one of those times.

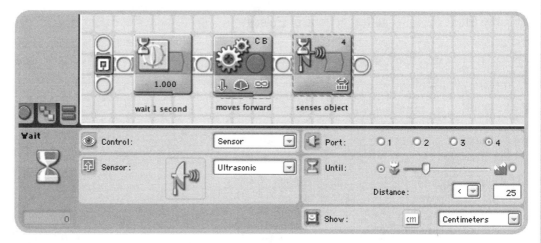

Figure 9-10: Now place a Wait block on the beam and configure it as shown here.

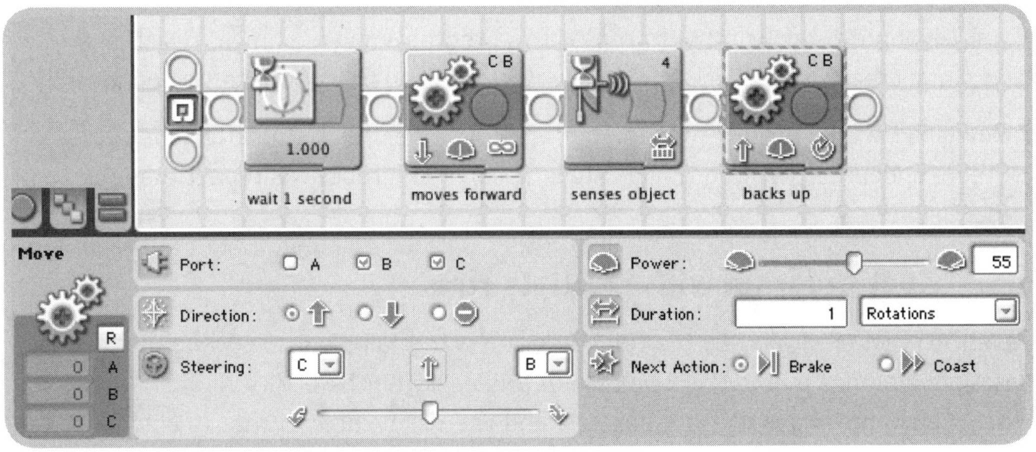

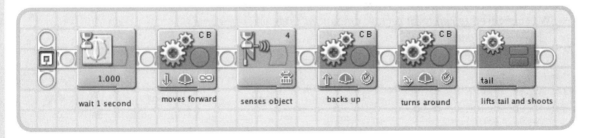

Figure 9-11: After the Ultrasonic Sensor detects an object located less than 25 cm in front of it, we want the robot to back up a bit (to keep from hitting the tail on anything).

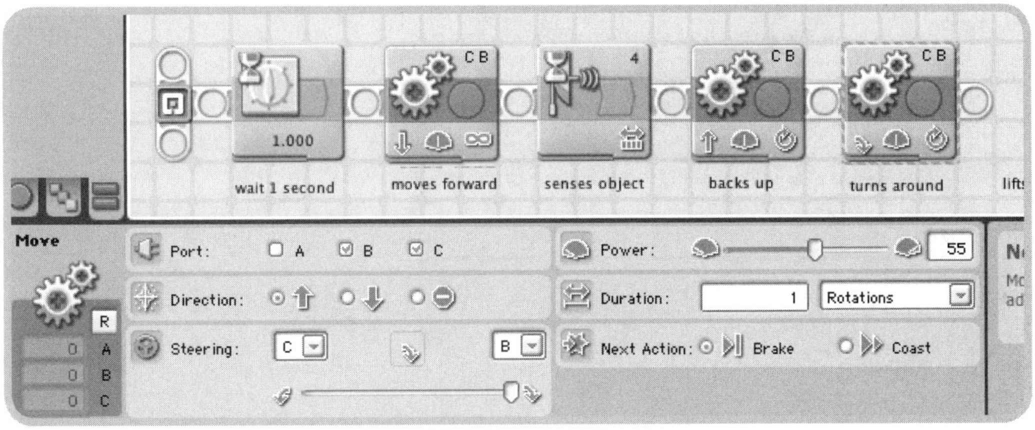

Figure 9-12: After backing up, we want Polecat to turn around, so we'll add a Move block, as configured here.

Figure 9-13: Finally, insert the tail My Block at the end of the beam. This means Polecat will lift its tail and shoot the dart when the robot turns around.

If you want your robot to repeat this program multiple times, select all the blocks and create a My Block named *skunk_turn*. Then you will have two options:

1 Download skunk_turn directly to your brick. The robot will stop after one cycle.
2 Create a loop for the program, as shown below. The robot will repeat the cycle multiple times.

final program with multiple runs

Save the main program and the final program with multiple runs, giving them brief, descriptive names that you will be able to find easily on the brick.

Now, download these separately and you will be able to choose between running Polecat through its paces once or many times.

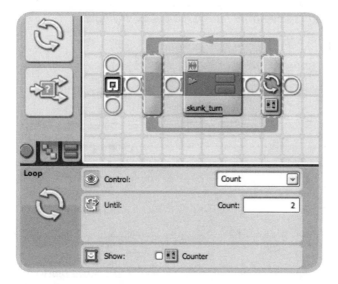

Figure 9-14: If you want Polecat to keep going after the dart is ejected, insert the skunk_turn My Block into a loop. A counting loop (as shown in the figure) or a loop configured to run for a certain amount of time will serve this purpose well.

strutter: an NXT peacock

This robot is a peacock who does his mating dance for everyone—and everything—he meets. Just like his namesake, he turns and flutters his tail feathers at any potential mate. Using some artificial feathers adds greatly to the effect and doesn't overtax the motor.

The pulley setup for the tail pieces is a bit tricky, but the programming is surprisingly simple, and the results are worth the effort.

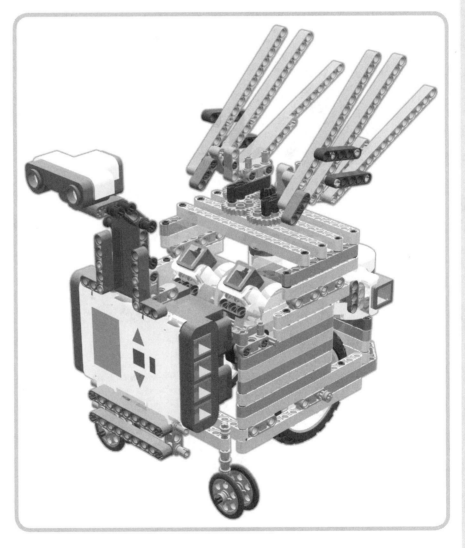

Figure 10-1: Strutter, the NXT Peacock!

building strutter

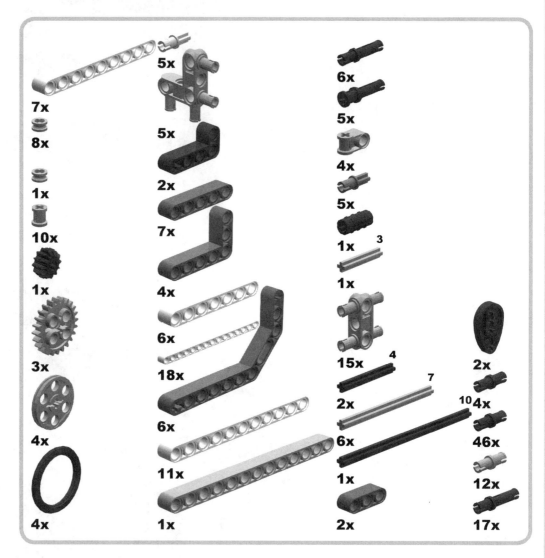

7x
8x
1x
10x
1x
3x
4x
4x

5x
5x
2x
7x
4x
6x
18x
6x
11x
1x

6x
5x
4x
5x
1x 3
1x
15x 4
2x 7
6x
1x
2x

2x
10 4x
46x
12x
17x

1x

2x

1x

2x

3x

* Non-LEGO part required: Fishing line
* Optional: Artificial peacock feathers, available at all major craft stores

front wheel (build two)

left wing

4

1x
3x
2x

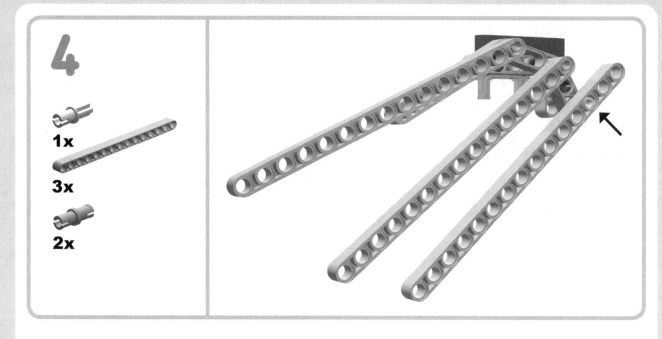

Use a yellow axle pin here.

5

2x
2x
4x

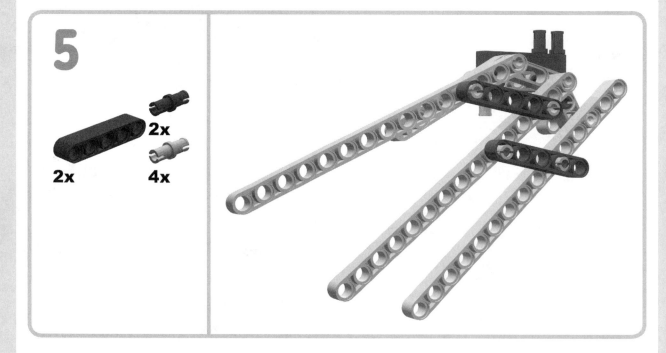

Use gray pins on the moving wing connections.

6

2x

1x

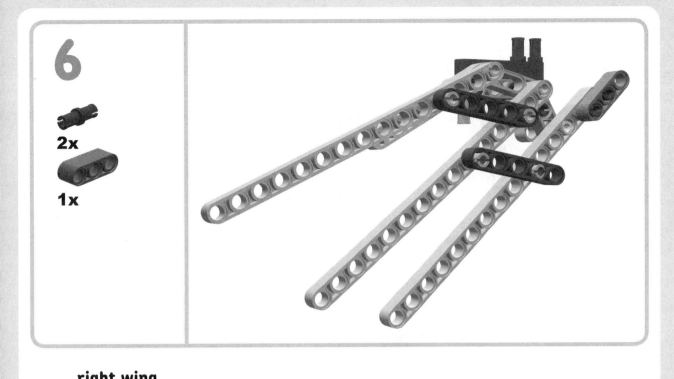

right wing

1

1x

1x

2

2x

1x

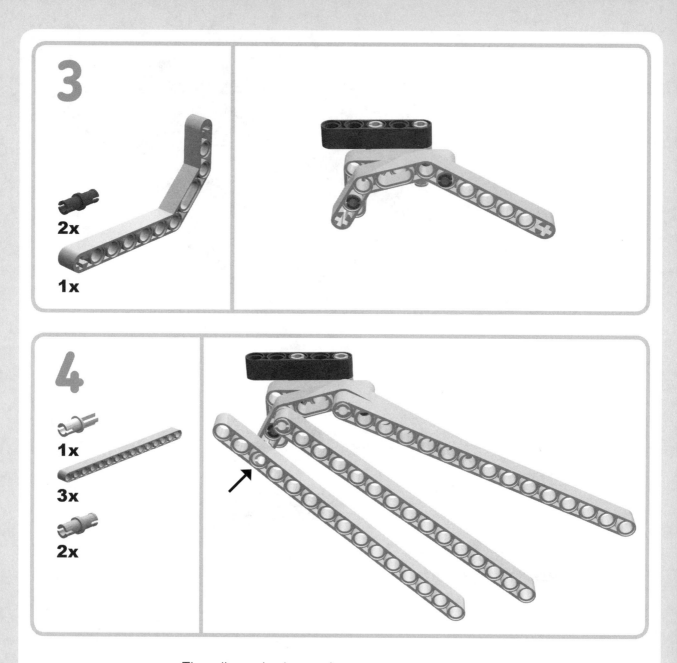

The yellow axle pin goes here.

5

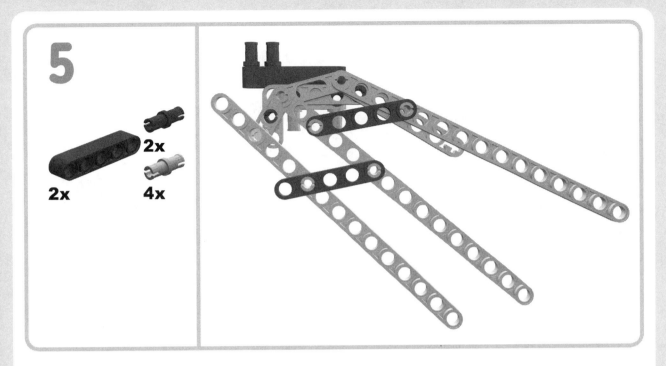

Use gray pins on the moving wing connections.

6

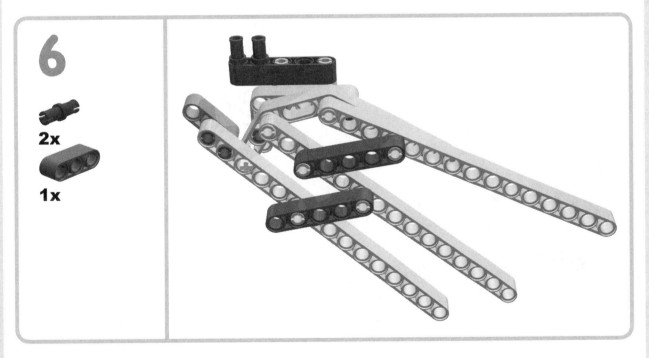

full assembly

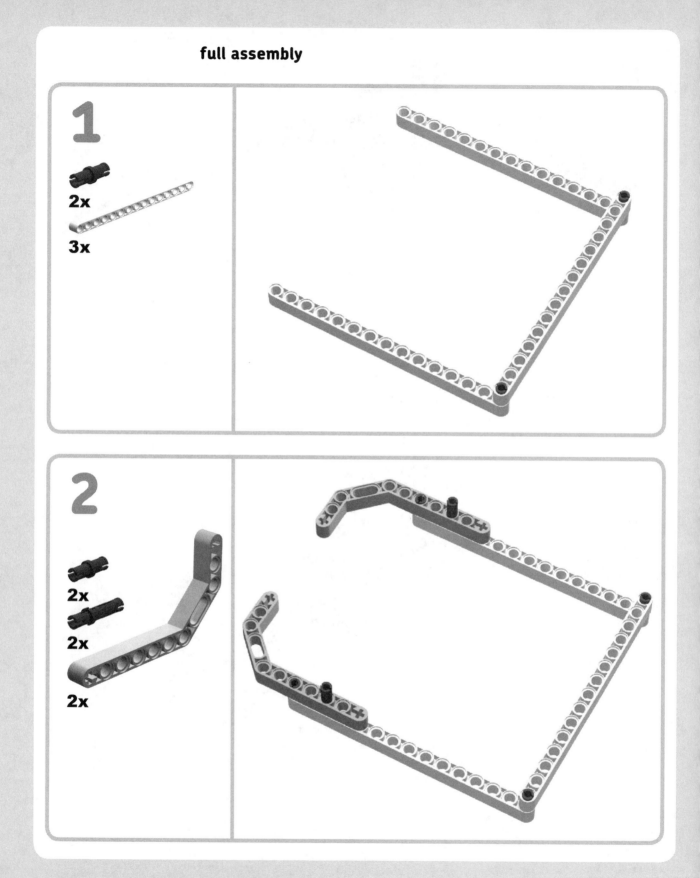

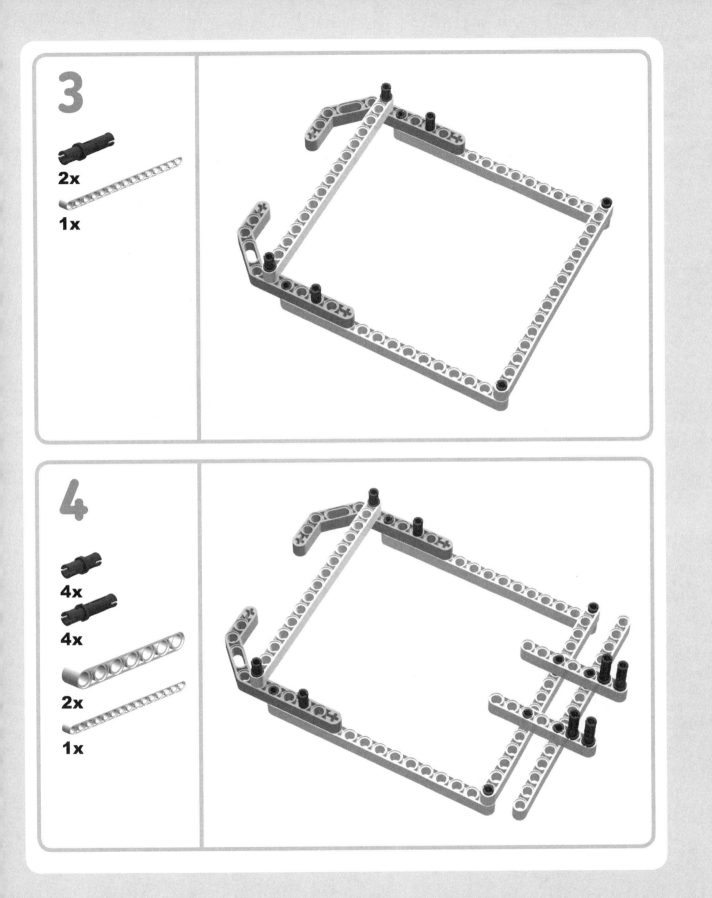

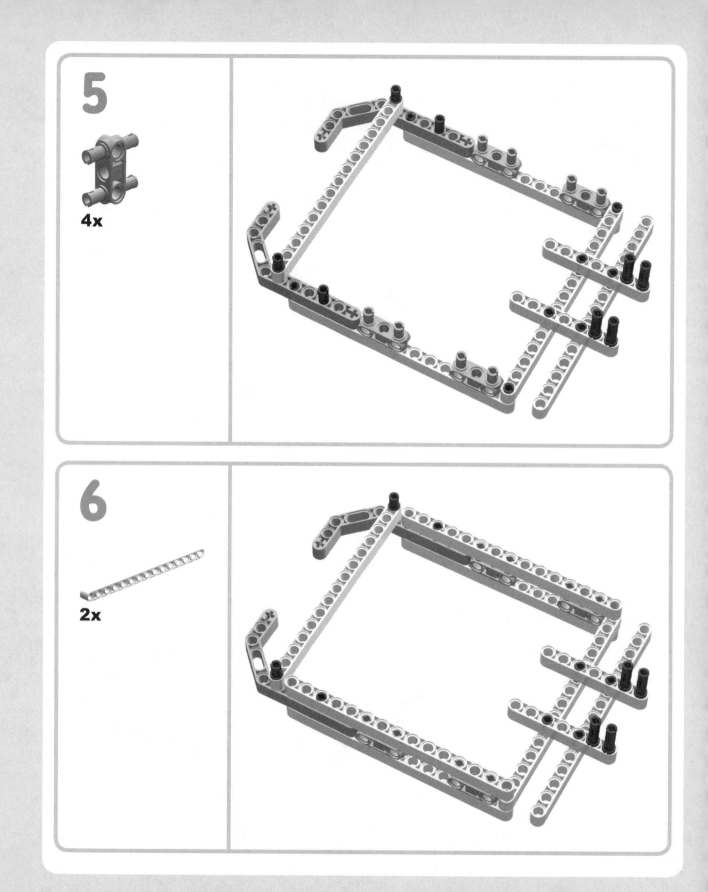

5

4x

6

2x

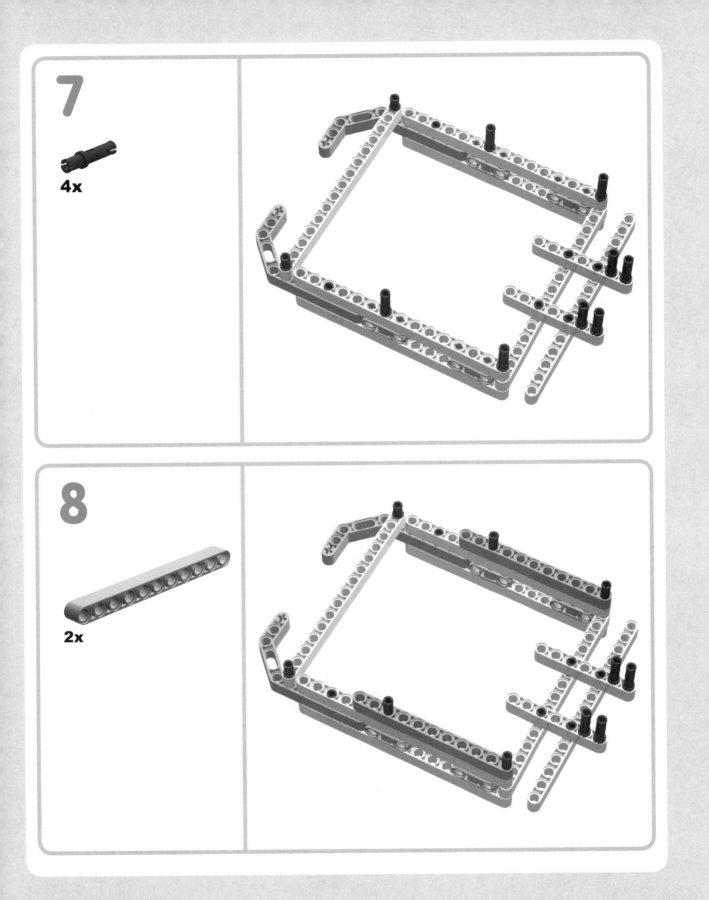

7

4x

8

2x

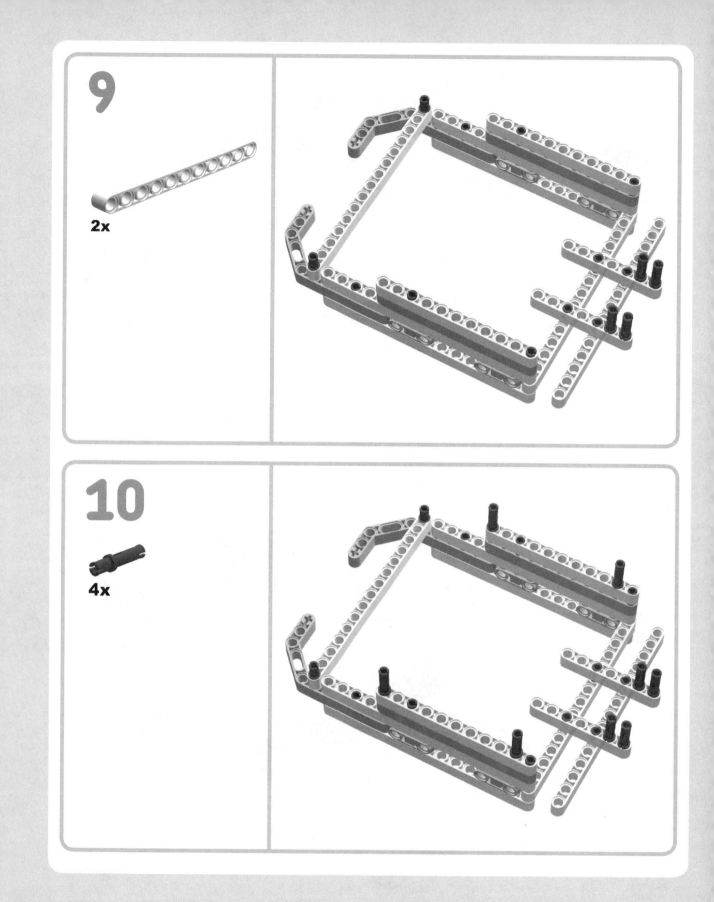

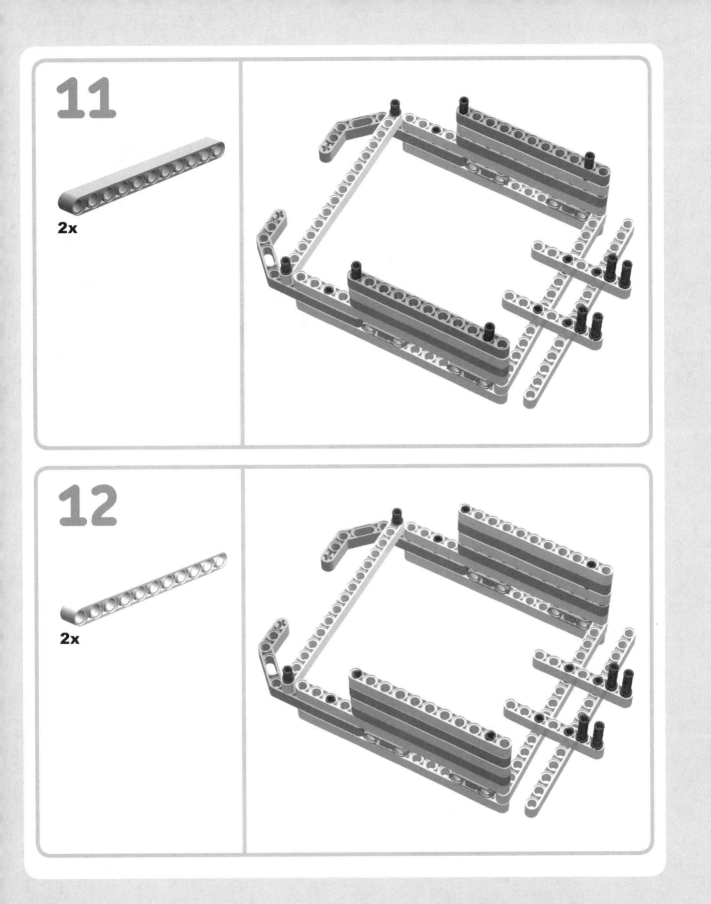

11

2x

12

2x

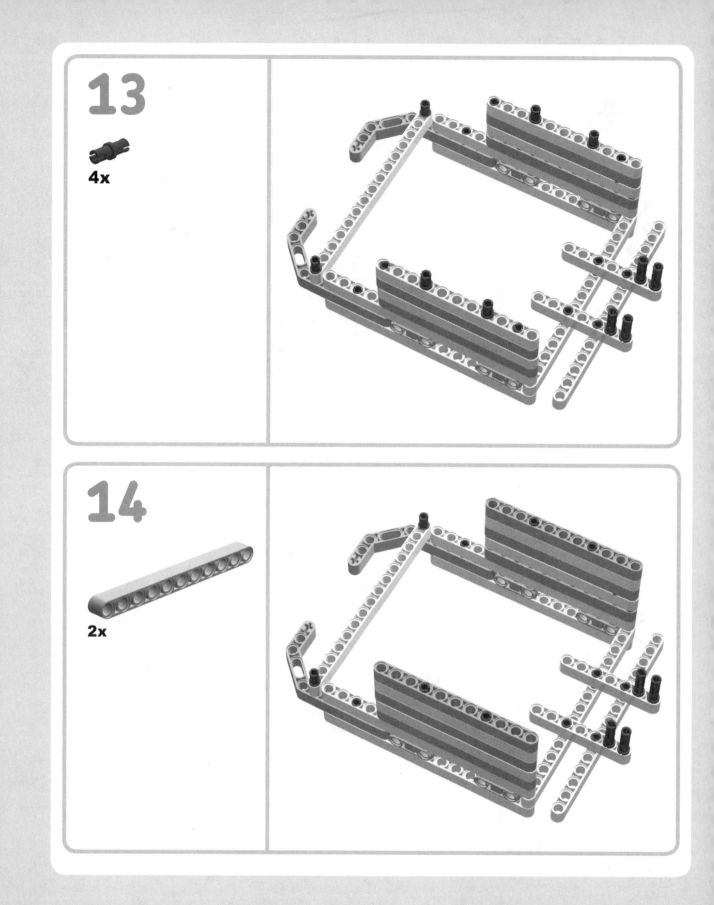

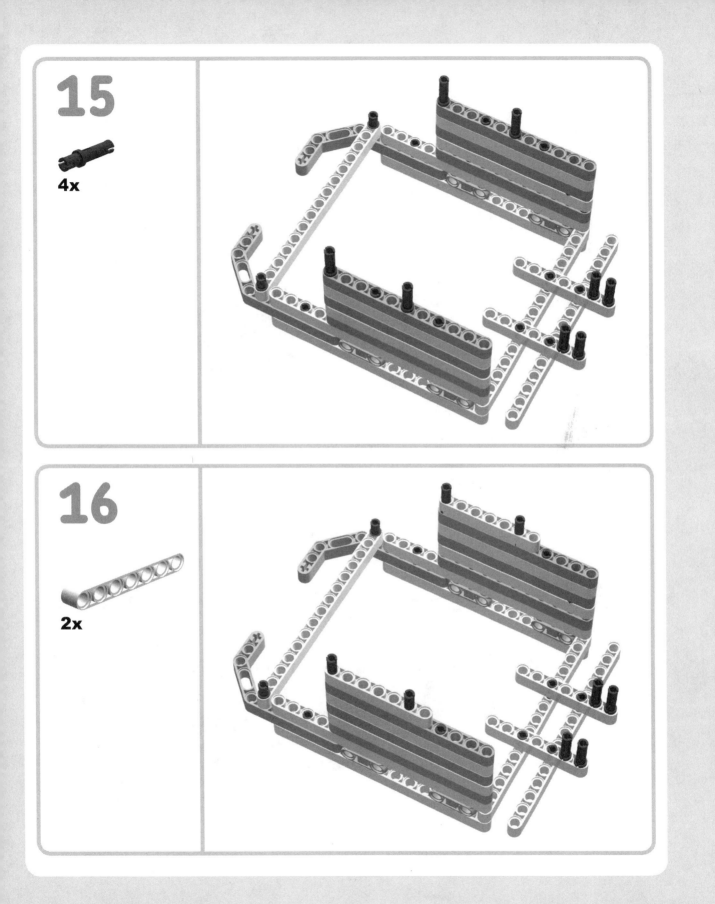

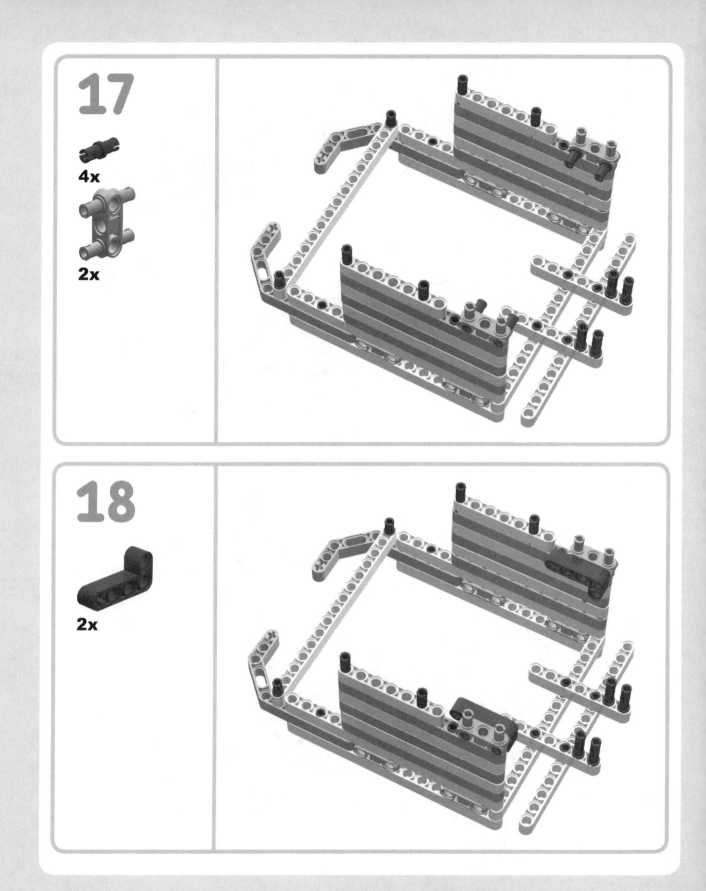

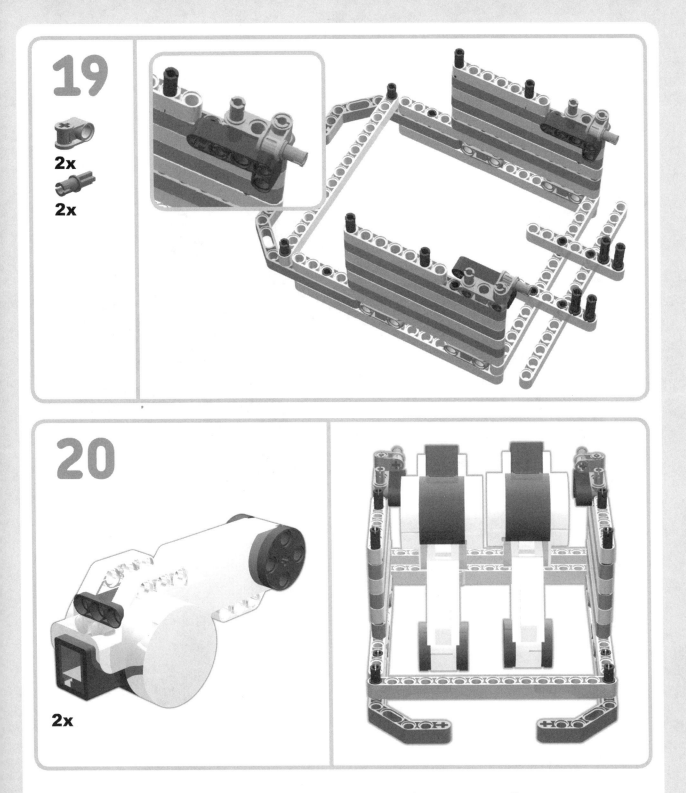

Step 20 is just to show you the positioning. In the next few steps, you will make the connections.

21

2x

7

2x

2x

2x

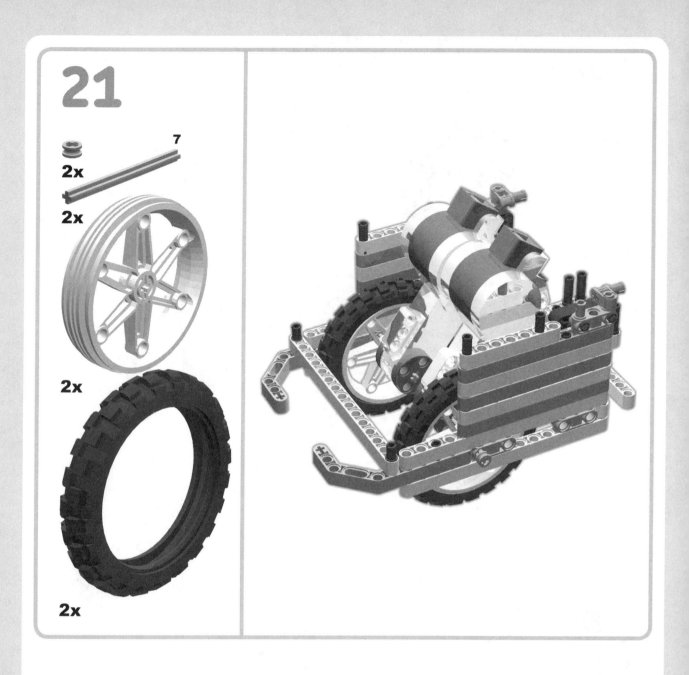

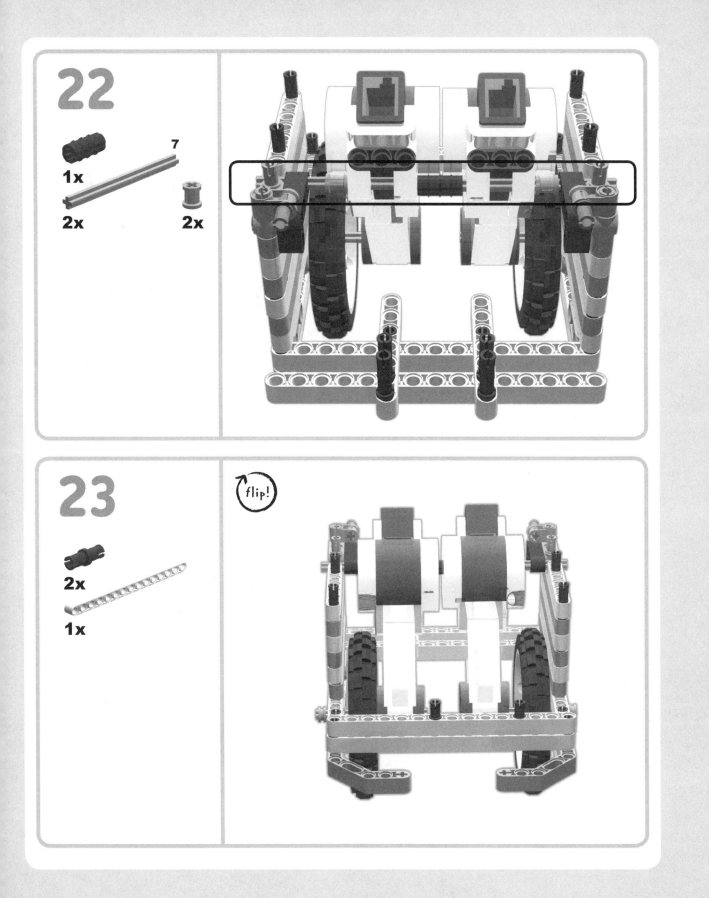

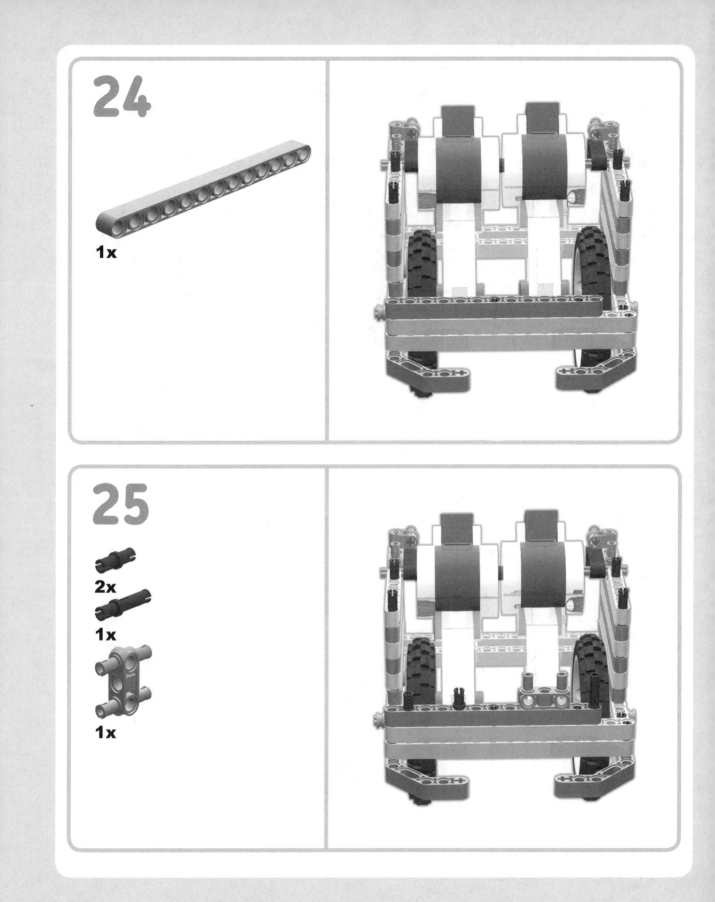

24

1x

25

2x

1x

1x

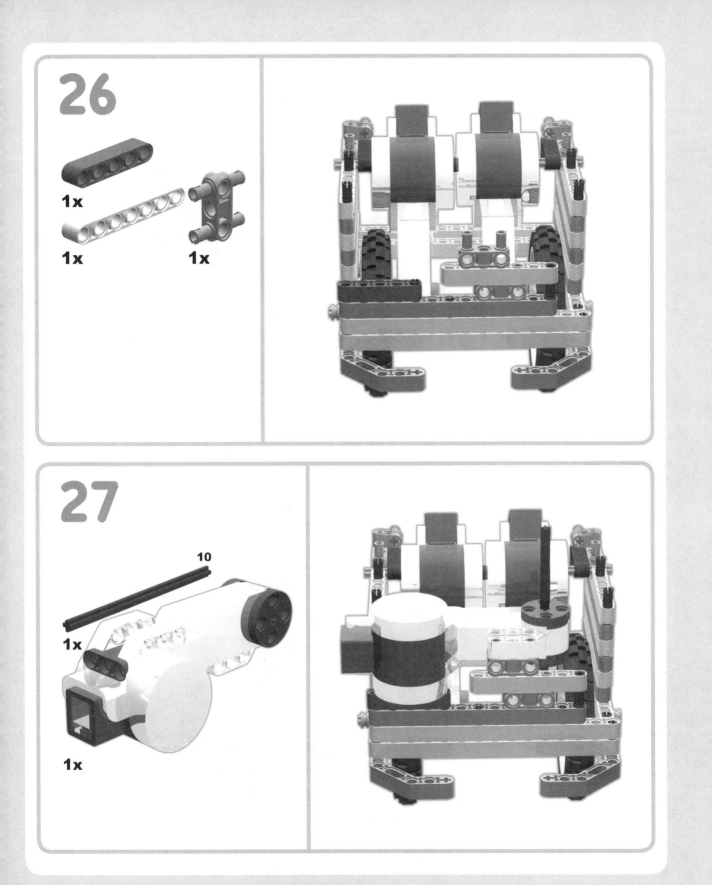

26

1x

1x

1x

27

10

1x

1x

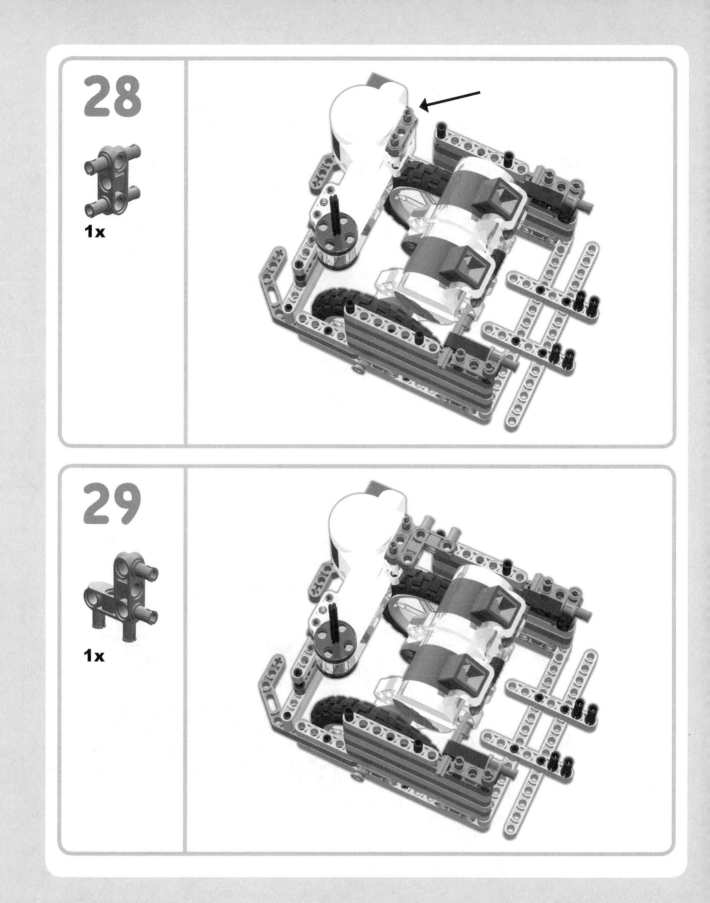

28

1x

29

1x

30

1x

1x

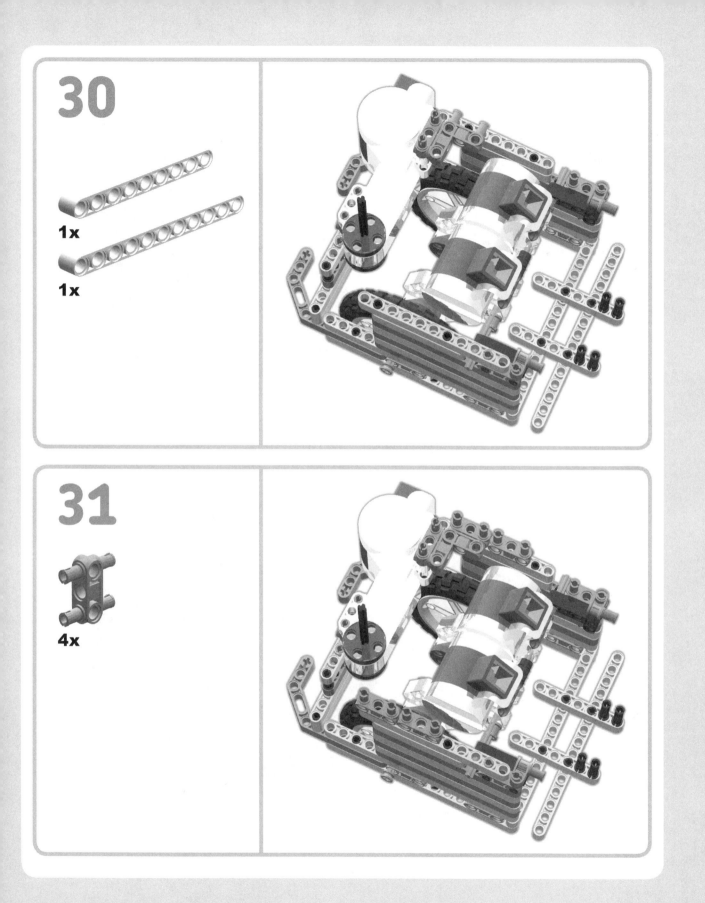

31

4x

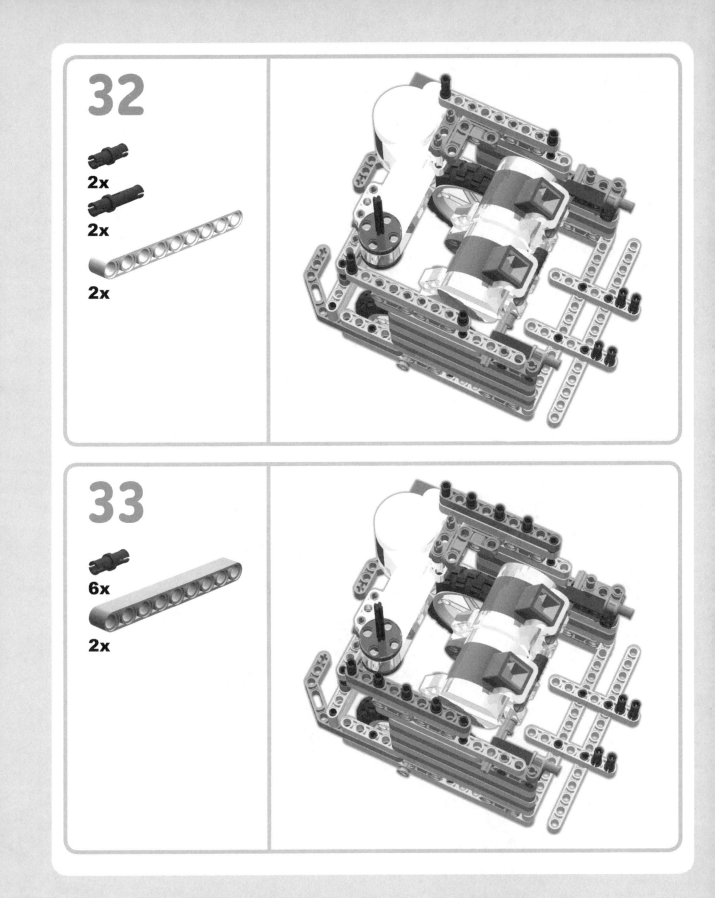

34

1x **3** **2x**

1x

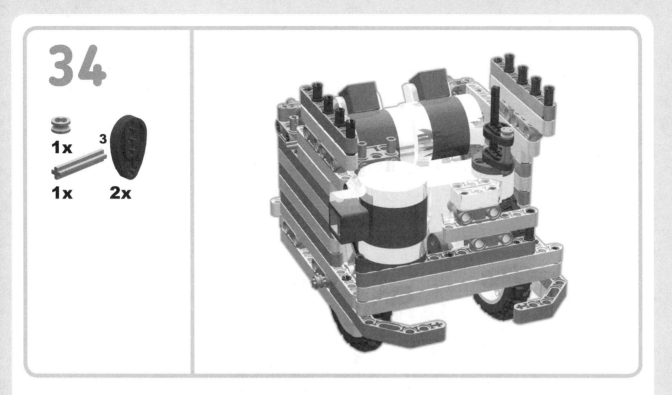

The small axle is in the first hole at the narrow end of the cams. The large axle goes through the first hole in the other end of the cams.

35

1x

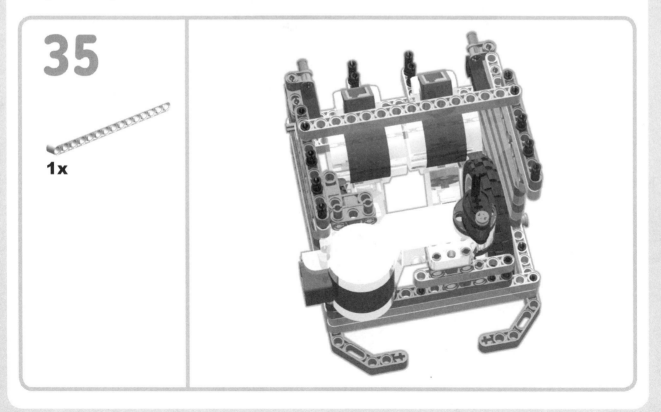

a pulley system for the wings #1

We're going to add the pulley as we go. First, take a 30-inch piece of fishing line (the least-stretchable kind). Tie the center of the line securely to the gray axle (pictured here).

After tying the center of the line to the axle, move the two ends front and left; then thread them through the center hole of the piece you placed in step 35. For now, leave the lines as they are—we'll finish this pulley later. You may eventually need to readjust the starting position of the axle subassembly in order to get maximum pull.

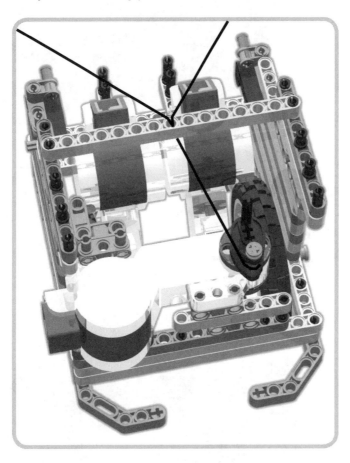

36

3x

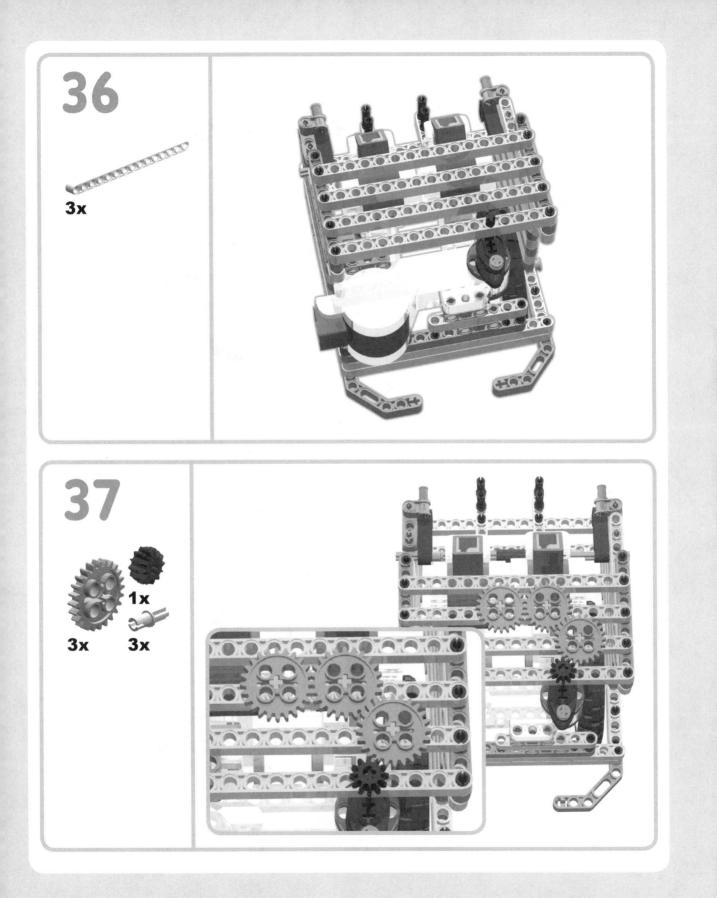

37

3x **1x** **3x**

38

1x

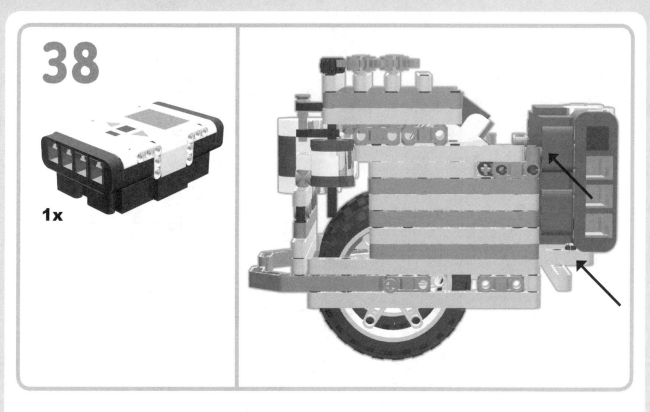

The NXT brick connects to Strutter's body at these locations on both sides.

39

2x

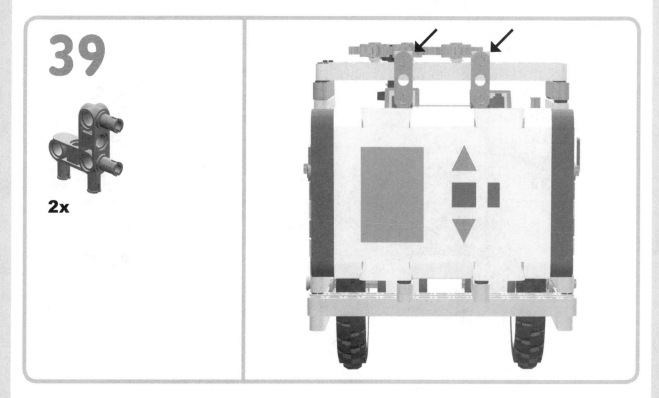

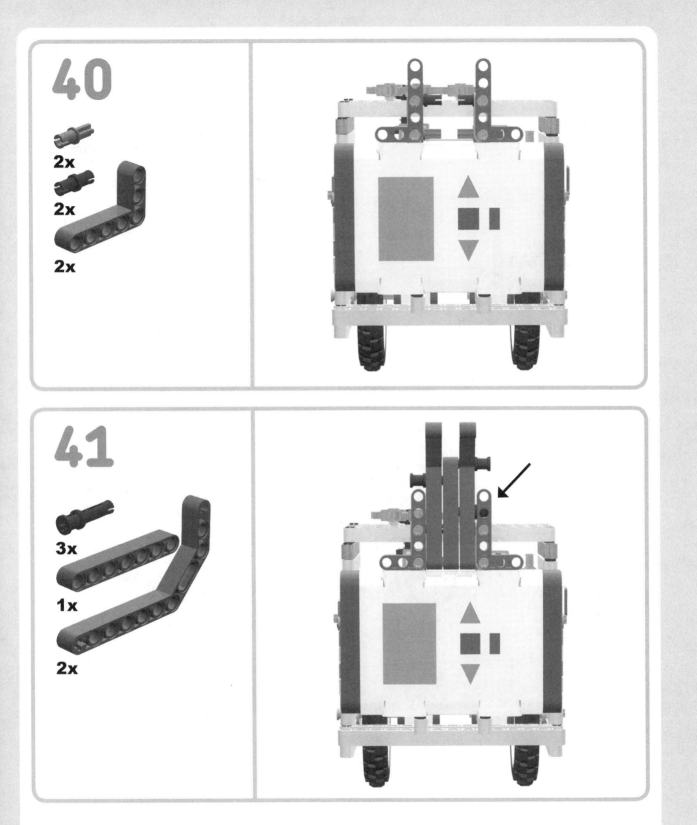

The third pin goes behind here.

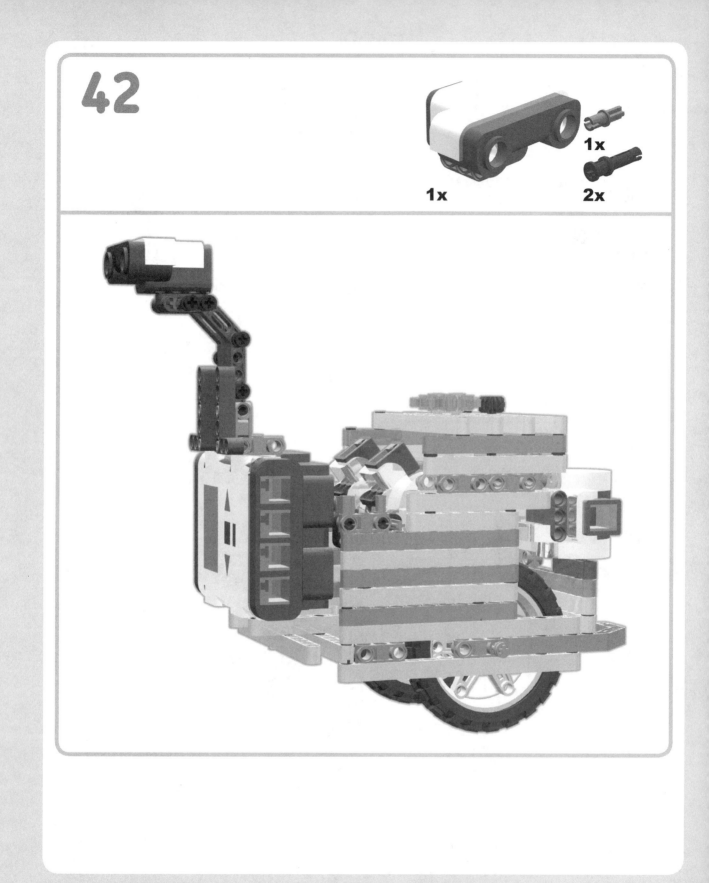

1x

1x

2x

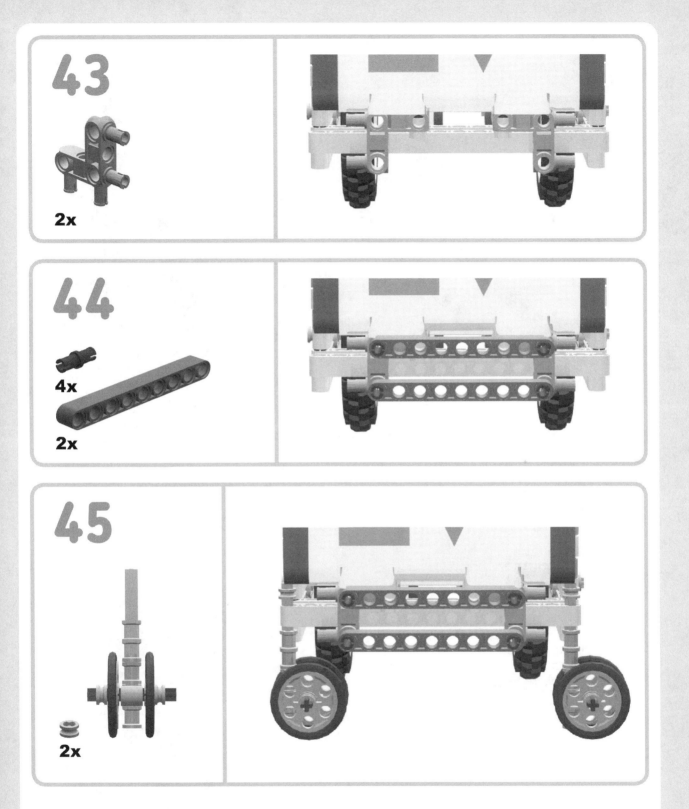

Put one wheel assembly and one half-bushing on each side.

46

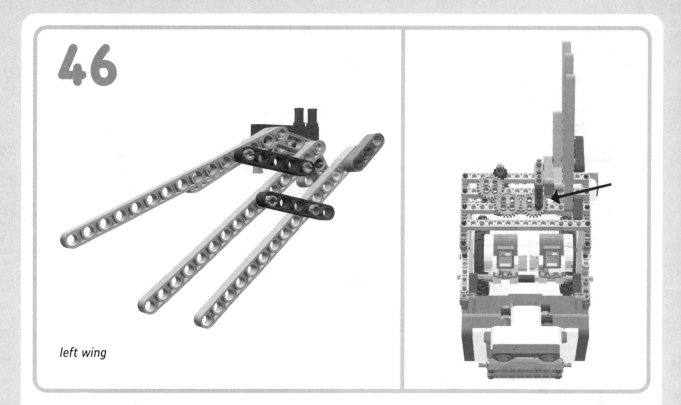

left wing

These black pins are inserted into the gears as shown.

47

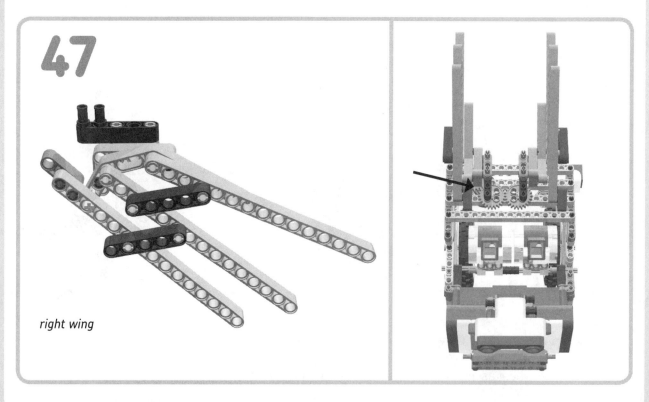

right wing

a pulley system for the wings #2

Take the two ends of the fishing line that you threaded through the center of the 15-hole beam after step 35 and attach each end of the line to the end of the 3-hole beam on each wing by tying it securely.

You do not want any slack in these pieces of line. If there's no slack, the wings will fan open and rotate around when the motor moves.

wiring connections

Connect the wheel motors to ports B and C. Connect the tail motor to port A. Connect the Ultrasonic Sensor to port 4.

troubleshooting tip

The pulley strings must be taut for the wings to work correctly. We found that turning the front of the wings slightly inward toward each other as you tie the fishing line to the tip of the wings was effective in helping us get the right amount of tension. (It pulls the string tighter when the wings fall back into place.)

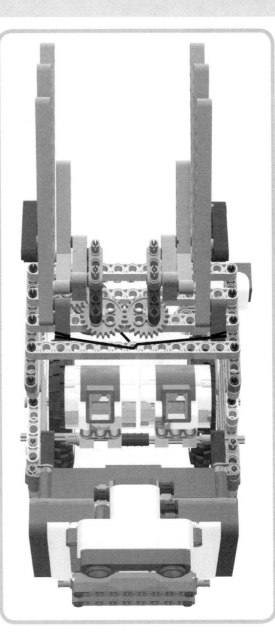

programming strutter

Let's begin with the creation of two My Blocks. The first will be Strutter's mating dance, and the second will tell Strutter when to do the dance.

my block #1: shake

First, let's get the wheels moving.

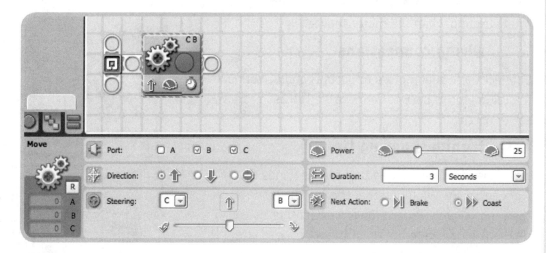

Figure 10-2: Place a Move block on the beam and configure it as shown here, with a power setting of 25. Set the duration for 3 seconds and the Next Action to Coast.

We are actually asking Strutter to back up here, but in our programming it is forward because of the position of the motors. This is something you need to keep in mind while programming. The direction you choose depends on the way you have installed your motors.

As Strutter moves slowly, we'll open up his wings.

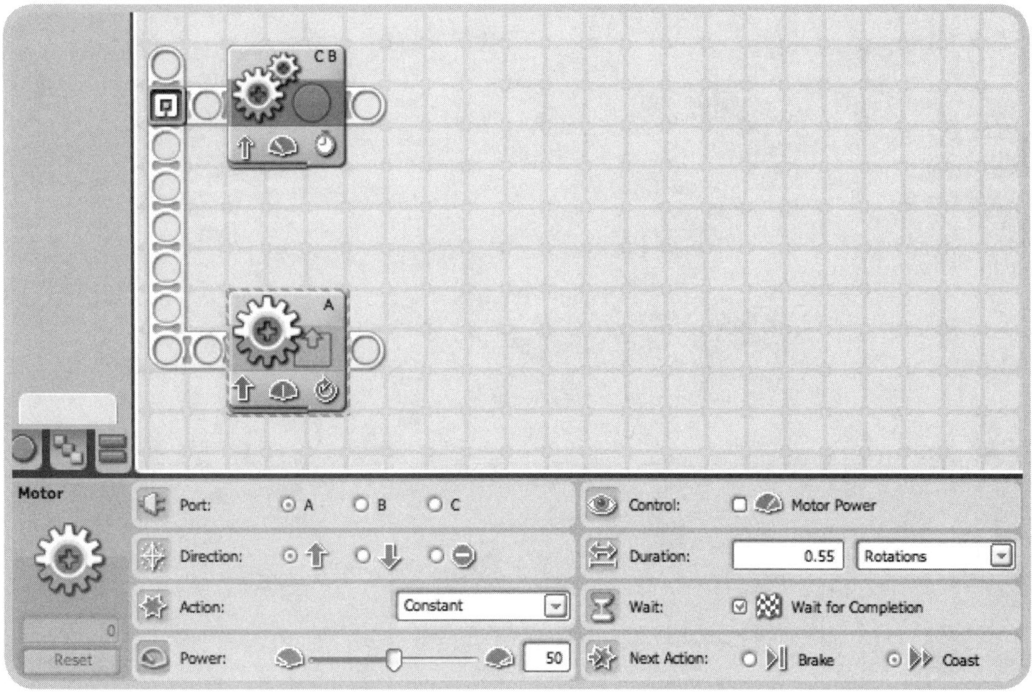

Figure 10-3: Place a Motor block on a parallel beam below the Move block. (Blocks on parallel beams perform their actions at the same time.) Port A should be configured for a power setting of 50 and a duration of 0.55 rotations in a forward direction, as shown here.

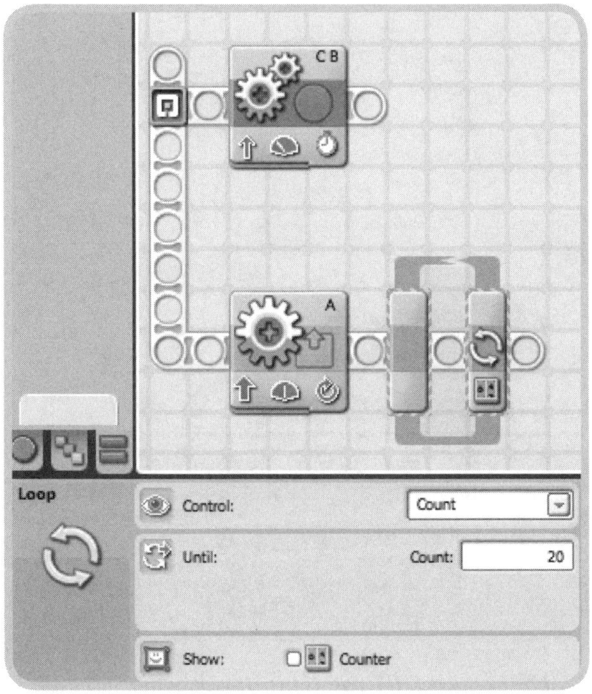

Figure 10-4: Next, drop a Loop block on the lower beam. Configure it to count to 20.

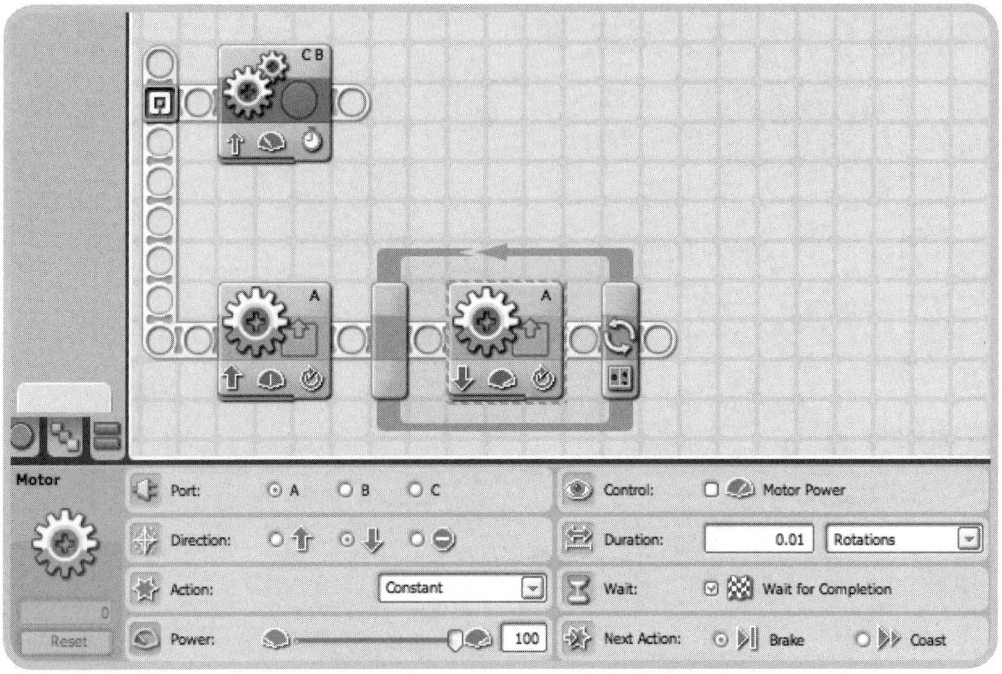

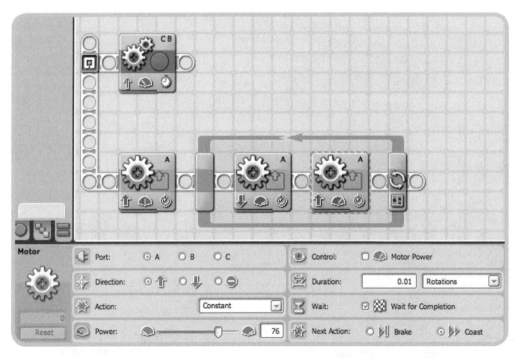

Figure 10-5: Place a Motor block into the loop, configured to go backward 0.01 rotations. The power is high because we'll make this motor work hard. Note that this motor is configured to brake.

Figure 10-6: Now we'll program motor A to go forward 0.01 rotations (this is when Strutter will flutter his wings). This motor is set to a lower power and to coast. See the configuration panel here.

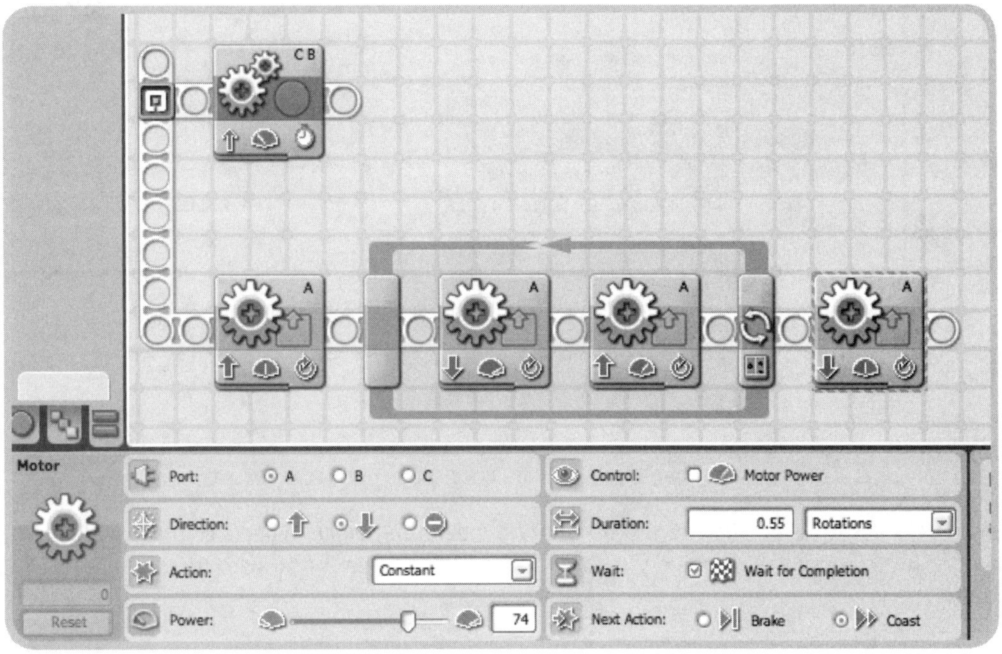

Figure 10-7: Finally, let's end the dance by causing the wings to swing back and down. We'll do that by configuring motor A to go back to its resting position.

Now save all these blocks as a My Block named *shake*. (Select all the blocks. Click the Create My Block button on the Toolbar, name the My Block, and give it a symbol so you can recognize it.)

my block #2: shakePlus

Now we'll program Strutter to respond to every solid object he meets.

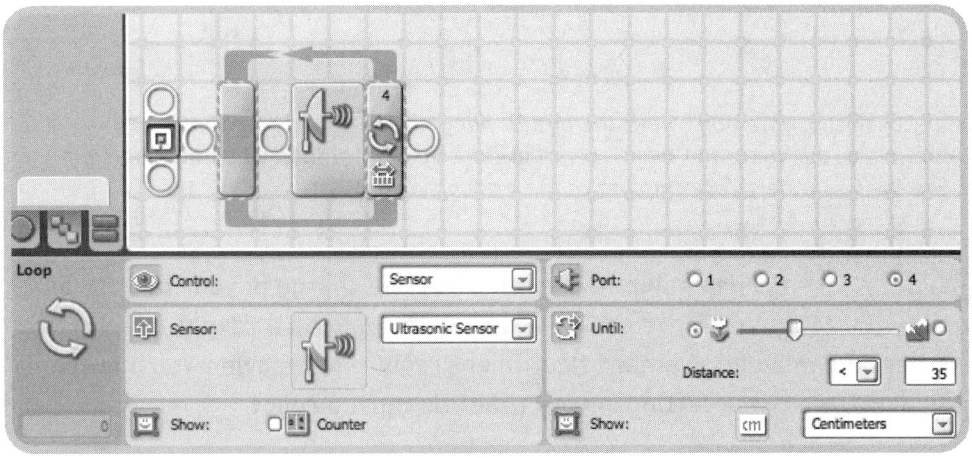

Figure 10-8: Open a new window and start with a Loop block configured as an Ultrasonic Sensor in port 4. It will respond to any object closer than 35 centimeters.

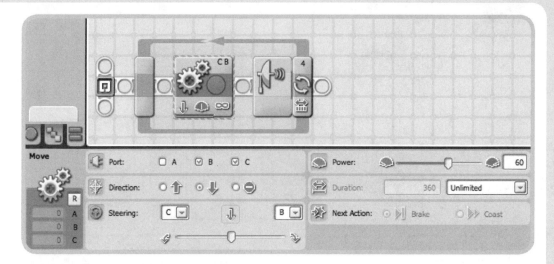

Figure 10-9: Here's what Strutter will do until he meets an object: Wheels B and C will move backward (which happens to move Strutter forward) at a power of 60 with an unlimited duration.

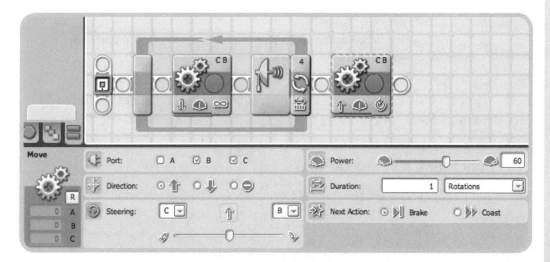

Figure 10-10: Because we don't want the bird to bang his tail against anything solid, we'll add a Move block, configured to make the robot back up one rotation when he meets an object. See the figure here for complete configuration settings.

NOTE Power settings depend on the strength of the charge in your battery and the surface the robot will move on. For example, moving on a carpet might need more power, and moving on a hard floor might require less power. You'll probably have to tinker with these settings every time you build a robot.

After Strutter has found an object, we want him to turn around and display his finery like a real peacock does.

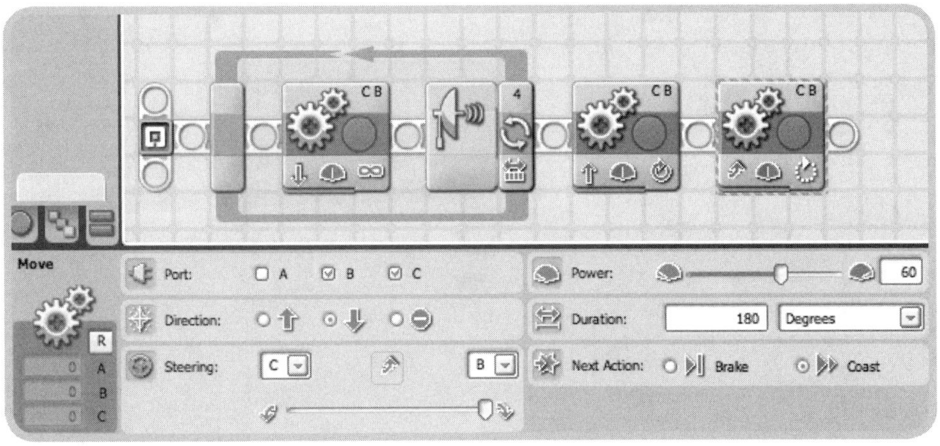

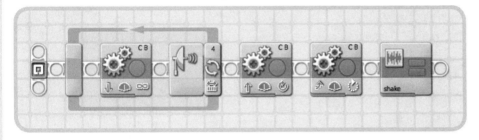

Figure 10-11: We'll program motors B and C to make a 180-degree turn backward. The steering slide is moved all the way over to B because we want a tight turn.

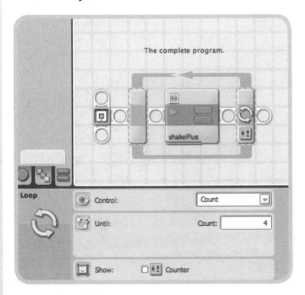

Figure 10-12: After he turns around, we want him to do his dance, so we'll insert the shake My Block.

Figure 10-13: My loop is configured to count four times and then stop the action. You can set it to count however many times you want, but I recommend setting a low number for your test runs.

Now select all the blocks and create a My Block named *shakePlus*. (Click the Create My Block button on the Toolbar, name the My Block, and give it a symbol so you can recognize it.)

the whole program

Because I don't want Strutter to move indefinitely, I'll use a Loop block to control how many times he repeats the dance. First, open a new window, place a Loop block on it and insert a shakePlus My Block in the loop.

Now you're ready to download the program and put Strutter into action.

enrichment ideas for teachers

As word has gotten out about this book, more and more teachers have expressed interest in using it in the classroom. While these suggestions for the classroom are hardly comprehensive, I hope they will be of assistance for teachers who are trying to get the most out of the NXT in their classrooms.

ultrasound

Here are some exploratory questions related to the NXT Ultrasonic Sensor:

* Test the accuracy of your robot's Ultrasonic Sensor by programming one of the robots to stop when the Ultrasonic Sensor threshold is triggered. (Be sure to select *Brake* in the lower-right part of the configuration panel for your motors.) Change the distance settings incrementally and chart how far from your solid object the robot stops. Next, change the power setting incrementally and chart the results. How does speed affect stopping time?
* Test the Ultrasonic Sensor on different surfaces. How do different textures affect accuracy? Why? How well does it detect mirrored or shiny surfaces? What about a carpeted surface? Can you think of some implications for your results?
* If you have two robots with Ultrasonic Sensors, do they still work correctly if they are close to each other? Test your hypothesis.

mathematical calculations in NXT-G programming

Programming the NXT offers opportunities to practice mathematical concepts. Here are some examples:

* How many degrees are in a motor's rotation? How can you convert a varying number of rotations to degrees? (See "Hopping Program #1: Random Hopping" on page 46.)

* Calculate the circumference of different tires. Use that information to program a defined path. How would the program change if you used larger or smaller tires?
* Calculate the relationship between speed and stopping distance using both the Brake and Coast settings.

CONVERTING INCHES TO CENTIMETERS

One common mistake made when programming the NXT is that when converting inches to centimeters, the programmer changes the number before changing the unit of measure. Here is an example of a problem you might pose:

> When programming distance for the NXT, the default unit of measure is inches. Let's say we want to set the distance to 5 centimeters. If we first change the number to 5 and then change the unit of measure to centimeters, the result will be 12.7 centimeters. On the other hand, if you first change the unit of measure to centimeters and then enter the number 5, the result will be 5 centimeters—just what you are looking for.

In addition to this being a lesson about accurate conversion, it also is a lesson in how important it is to do things in the correct order.

simple machines

The simple machines being used in these robots include gears, levers, pulleys, and wheels and axles.

Ask students to identify the simple machines used in each robot. Then have them identify the work accomplished by each one.

ideas for investigation

Here are some suggested investigations for using the NXT to practice the scientific method.

investigation #1: what is the best leg length for your robot?

This activity can be done with any walking robot in this book.

Have teams design their own robotic legs; then test each design on one robot to see what effect each design change has on the action of that robot. Do the same for different robots.

investigation #2: how does leg design affect the center of gravity?

Walking robots offer an unparalleled opportunity to experiment with the center of gravity. Have students modify the shape and length of their robot's legs; then test each design. Chart the results. What are the characteristics of the legs that function best (keeping the robot stable and upright)?

Now think about real-world artificial legs. Do some research into the way artificial legs move. Can you use the same mechanism found in prosthetic legs on a robot? Why or why not?

investigation #3: how does the method of programming a robotic turn affect the result?

There is more than one way to program a turn in a robot.

Using a wheeled robot with two motors pointed in the same direction (as in Strutter the peacock), compare the results when you program just one side to move for a turn versus programing each side to rotate in opposite directions (for example, the left motor moves forward and the right motor moves backward).

Program for 90-, 180- and 360-degree turns and then compare the results. Come up with a strategy for programming the most accurate turn.

investigation #4: how can you make a specific robot go exactly where you want it to go?

Select a specific distance that you want your wheeled robot to move and then stop, or pick a target for a robot to stop on. Write a program to move your robot to this specification. How does the size of the wheels affect the way you write your program?

Now look at Strutter. Do you need to consider the circumference of the front wheels when you are writing your code?

Compare the response of Polecat the skunk to Strutter. Their wheels are in opposite positions (small wheels in front on Strutter, small wheels in back on Polecat). How does that affect the results? Are they the same? If not, why not?

investigation #5: how does weight distribution affect function?

Polecat is an excellent robot to use to study this question. For example, you will find that this robot acts differently if you add fur to its tail. You might also try shifting the brick back a notch, adding more weight to the back, to see how it affects the robot. What does this tell you about traction and vehicle movement?

investigation #6: how can we accurately program a turn in a specific walking robot?

Using a walking robot, do the following:

* Calculate the minimum distance required for the robot to make a turn. How does this number change when you modify the turning method?
* Increase the speed of one side. (This requires using a different motor for each side.)
* Configure opposite directions for each side. (This also requires using a different motor for each side.)
* What happens when you change the settings in various ways?
* Use degrees, time, and rotations for the Duration setting. What happens?
* Change the power levels.

Finally, chart and compare all your data. How do the observed results compare with the settings in your program?

investigation #7: which string is best for use on pulleys in robots such as pygmy and strutter?

Collect different kinds of string such as dental floss (various types), fishing line, strings used for bead work, yarn, twine, cotton string, thread, and so on. Be creative. Devise methods to test a string for the following:

Bulk	Does it fit where you want to use it?
Ease of use	Is it easy to tie?
Elasticity	Does it stretch too much—or not enough?
Flexibility	Does it bend as required by your project?
Friction	Can you tie it so that it stays tied?
Strength	Does it break too easily?
Texture	Will it get caught in the mechanisms?

Record and chart the results of each test; then decide which strings are the most suitable.

How can you apply what you've learned to real-world situations? For example, if you tie your shoes with different materials, which one works best, and why? What are the characteristics of a successful shoestring?

using your findings

Based on what your students have discovered, devise a competition that requires them to apply the knowledge they've gained in their research. Some examples include the following:

* Perform time trials for robots using the leg designs you have created. How long does the robot take to move from one point to another without breaking down? Make this a team competition, allowing each team to use a leg design of its choice but the same robot design.
* Put a target on the floor. Set a starting point and see whether students can program their robots to end up exactly in the target's bull's-eye.
* Create a winding path or obstacle course on the floor. Using a robot with an Ultrasonic Sensor, write a program that causes the robot to travel through the maze. Award extra points for getting through the maze with the least amount of contact with the sides. Use time trials for speed.
* Create a list of specific actions for a robot. Award points for successfully programming a robot to re-create each action on the list. These could be actions such as making a perfect 90- or 360-degree turn, traveling a specific distance, getting a walking robot to walk a certain distance in a straight line (without the gears grinding or falling off), and so on.

the arts

Rarely does anyone mention the artistic potential for the NXT, but that is what I enjoy most about it. It allows me to create art that responds to its environment with sound and motion. Here are some suggestions for how you might merge NXT robotic building with the arts:

* Select simple stories and create robots to represent their characters in your own LEGO robotic video production. If you don't have equipment for this, see if parents or local businesses will help.

* Write and create your own stories, faux news reports, commercials, or public service announcements using your robots as characters. (We had great fun creating a video that pitted a human curler against a robotic curler. The robot won!)
* Have everyone take the same basic robot and dress it up as an original character. Ask students to create a history for the character: place and date of birth, occupation, likes and dislikes, accomplishments, and so on.

This kind of activity is great for engaging everyone in the class. The various interests and talents make for varied—and wonderful—results.

B

troubleshooting

No matter how good the instructions are, it's a challenge to build any robot just right the first time. In fact, I can think of only one or two times when my first test of a walking robot was completely successful. To that end, I hope that these troubleshooting tips will prevent any serious frustration.

Before I talk about fixing problems, let me emphasize some ways you can prevent problems in the first place:

* In the walking robots, all the legs need to be *exactly* the same size.
* Gears must always be lined up exactly.
* Make sure you are using the correct parts. If you use parts that are slightly different, your model might act differently.

You might need to change power or other settings, depending on your battery strength, walking surface, and so on. Some models, such as Sandy the camel, move better at a slower pace, while others, such as Pygmy the elephant, will march right along at higher speeds.

NOTE **You will see a definite change in the function of your robots as your battery weakens. This is particularly obvious with the jumping robots.**

are your gears grinding and pulling the robot apart?

The pairs of gears running each leg must be oriented *exactly* the same way, as shown in Figure B-1. If they are even slightly out of alignment, the gears will grind and your robot will not walk correctly.

To fix grinding gears:

1 Observe the moving robot to see which side is grinding. (Hold the robot up in the air to observe this.)
2 If just one side seems to be a problem, remove the legs from that side. (You might need to remove all the legs to follow these problem-solving procedures.)

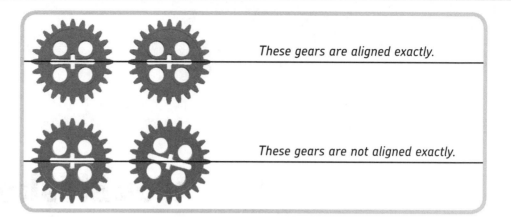

These gears are aligned exactly.

These gears are not aligned exactly.

Figure B-1: Even this slight difference will keep a robot from walking correctly.

3 Remove the driving gear (the gear connected directly to the motor axle) so that the other gears can move freely; then test the movement of the gears. If they are all moving freely, go to the next step. If they aren't moving freely, observe what is getting in the way and fix the problem. Is a motor axle sticking out too far? Is a gear rubbing against something? Did you use the wrong kind of pin (blue instead of yellow, or black instead of gray)?

4 Check the legs to make sure they are built correctly. Is the correct part of the foot under the robot?

5 Put the driving gear back in place and, if possible, use a bent connector (as shown in Figure B-2) to hand test the freedom of motion of the gears. Turn the handle in both directions. If the gears aren't moving freely, observe what is obstructing the movement and eliminate it. This is often a gear rubbing against another object, such as a pin that isn't pushed in far enough.

6 Make sure that each pair of gears (the two gears supporting one leg) line up *exactly*. It might not be easy to line the gears up precisely, but it is absolutely necessary.

7 Attach the legs, one at a time. Then, using the handle on the driving gear, test the movement of the gears—forward and backward. If everything moves freely when all four legs are reattached, run your program again.

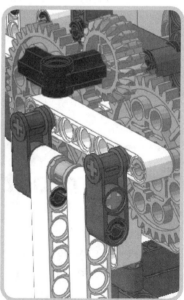

Figure B-2: Driving gear with connector handle attached

what if your robot walks in place, instead of moving forward?

If your robot walks in place, you have probably not arranged the gears and legs as described in the building instructions.

Look at the example from LEGOsaurus, shown in Figure B-3.

Both sides look exactly the same *as you face them*—top-left position on the left, and bottom-right position on the right. The basic principle is that the front left leg should be in the same position as the back right leg, and the front right leg should be in the same position as the back left leg. If you don't follow that basic principle, you'll find that your walking robot merely moves in place—or sort of rolls up and down, like it is riding the waves.

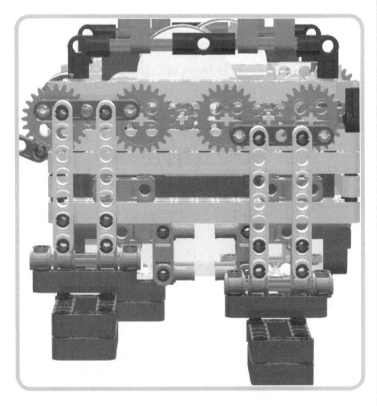

Figure B-3: LEGOsaurus's legs are positioned like this.

does your robot seem to be ignoring your programming directions?

The best way to find a programming problem is to break it down into small sections and then add more steps gradually. For instance, if your program has a number of My Blocks, download the blocks separately to the brick and test each one. If that doesn't solve the problem, break down your My Blocks into smaller units and increase from there. If you work your way up, you should be able to find the trouble spot (or spots).

Another great troubleshooting tip (thank you, Jim Kelly) is to place sounds at key action points in your program. That way you'll be able to hear where your program is being derailed.

SOMETIMES IT HAS NOTHING TO DO WITH THE PROGRAMMING

When I assembled my Polecat for a final test, I was mystified when it refused to move forward as the first program block directed. I rewrote the program and stripped it down to its absolute minimum, but nothing worked. It was very frustrating because it had worked perfectly the last time I built it.

Then, quite by accident, I pressed the orange button on the NXT brick and started the robot as I held it in the air. To my surprise, the wheels were turning forward.

When I set the robot down, it immediately stopped moving forward. I still couldn't see anything to prevent the robot from moving forward. My only solution appeared to lie in moving, removing, or changing every part that was somehow involved with moving the wheels forward.

I had never seen a bushing prevent an axle from turning before, let alone stop it from turning a wheel in just one direction. Just the same, removing the bushings from the wheel axles solved the problem completely.

whatever you do . . .

Be patient. These robots will work as long as you follow the instructions provided.

web resources for the NXT

The Internet offers a lot of information about the LEGO MINDSTORMS NXT. These sites will be helpful to anyone using this book:

http://www.thenxtzoo.com/

> This companion site for *The LEGO MINDSTORMS NXT Zoo* offers the following:
>
> * Instructions for creating custom sounds for an NXT robot
> * Links to sources of NXT parts
> * Updates for the zoo models
> * And more . . .

http://thenxtstep.com/smf/index.php

> This is the forum monitored by members of The NXT STEP blog. If you search for a topic you want information about, such as *RSO sound*, you'll find a variety of helpful discussions—or you can start your own topic with your specific question. The forum monitors are extremely well informed and helpful (as are many of the 30,000 people who visit every month), so take advantage of them. Feel free to ask questions—just be sure that you include as much information as possible. (Photos of problem assemblies are particularly helpful.)

http://mindstorms.lego.com/nxtlog/default.aspx

> This LEGO site is a place for NXT designers to share their creations. If you search for sound files, you will discover that many designers have also shared their favorite ways to create custom sounds.

> If you want to experience buying LEGO parts from around the world (eBay-style), check out these two sites:

http://www.bricklink.com/storeList.asp

> If you love LEGO building sets, spending time on this site is like being a kid in a candy store. You can find almost any LEGO part in colors you never knew existed, from sellers next door or around the globe. A few items are rare and rather pricey, but most parts are very reasonable.

http://www.peeron.com/

This is the ultimate resource site for LEGO parts. You can find complete parts lists, including part names and numbers, for every kit LEGO has ever made. For some reason, part names and numbers on this site may be different from those used by LEGO; but if you're persistent, you can usually find what you need. This site is particularly useful if you are considering buying a TECHNIC set to supplement your parts stash.

Here are two well-monitored, child-friendly NXT blogs:

The NXT STEP   *http://thenxtstep.com/*

nxtasy.org   *http://www.nxtasy.org/*

The following websites may be helpful for those interested in creating building instructions for their own NXT creations.

http://www.ldraw.org/

This is the most comprehensive site about CAD drawings for the NXT. It's a volunteer-run site, full of outstanding freeware. The building instructions in this book were created with freeware I learned about on this site (MLCad, POV-Ray, and LPub). I have to warn you that the software is challenging to learn and use—especially since there is little documentation for any of it.

http://ldd.lego.com/

This is the home of LEGO Digital Designer. You can draw NXT designs here, but you will only find the parts from the NXT Base Set, and you cannot order parts to build your design.

http://sketchup.google.com/

Someone has created some NXT parts for this Google drawing program, but you will not find all of the parts. My experience with Sketchup is that it is tricky to learn.

index

colophon

The LEGO MINDSTORMS NXT Zoo! was laid out in Adobe InDesign. The font is Chevin.

The book was printed and bound at Malloy Incorporated in Ann Arbor, Michigan. The paper is Glatfelter Thor 50# Smooth. The book uses a RepKover binding, which allows it to lay flat when open.

The Unofficial LEGO® MINDSTORMS® NXT Inventor's Guide
by DAVID J. PERDUE

The Unofficial LEGO MINDSTORMS NXT Inventor's Guide will teach you how to successfully plan, construct, and program robots using the MINDSTORMS NXT set, the powerful robotics kit designed by LEGO. This book begins by introducing you to the NXT set and discussing each of its elements in detail. Once you are familiar with the beams, gears, sensors, and cables that make up the NXT set, the author offers practical advice that will help you plan, design, and build robust and entertaining robots. The book goes on to cover the NXT-G programming environment, providing code examples and programming insights along the way. Rounding out the book are step-by-step instructions for building, programming, and testing six complete robots that require only the parts in the NXT set; an NXT piece library; and an NXT-G Glossary.

OCTOBER 2007, 320 PP., $29.95 ($32.95 CDN)
ISBN 978-1-59327-154-1

The LEGO® MINDSTORMS® NXT Idea Book
Design, Invent, and Build
by MARTIJN BOOGAARTS, JONATHAN A. DAUDELIN, BRIAN L. DAVIS, JIM KELLY, DAVID LEVY, LOU MORRIS, FAY RHODES, RICK RHODES, MATTHIAS PAUL SCHOLZ, CHRISTOPHER R. SMITH, *and* ROB TOROK

With chapters on programming and design, CAD-style drawings, and an abundance of screenshots, *The LEGO MINDSTORMS NXT Idea Book* makes it easy for readers to master the LEGO MINDSTORMS NXT kit and build the eight example robots. Readers learn about the NXT parts (beams, axles, gears, and so on) and how to combine them to build and program working robots like a slot machine (complete with flashing lights and a lever), a black-and-white scanner, and a robot DJ. Chapters cover using the NXT programming language (NXT-G) as well as troubleshooting software, sensors, Bluetooth, and even how to create an NXT remote control. LEGO fans of all ages will find this book an ideal jumping-off point for doing more with the NXT kit.

SEPTEMBER 2007, 368 PP., $29.95 ($35.95 CDN)
ISBN 978-1-59327-150-3

Forbidden LEGO®
Build the Models Your Parents Warned You Against!
by ULRIK PILEGAARD *and* MIKE DOOLEY

Written by a former master LEGO designer and a former LEGO project manager, this full-color book showcases projects that break the LEGO Group's rules for building with LEGO bricks—rules against building projects that fire projectiles, require cutting or gluing bricks, or use nonstandard parts. Many of these are back-room projects that LEGO's master designers build under the LEGO radar, just to have fun. Learn how to build a catapult that shoots M&Ms, a gun that fires LEGO beams, a continuous-fire ping-pong ball launcher, and more! Tips and tricks will give you ideas for inventing your own creative model designs.

AUGUST 2007, 192 PP. *full color*, $24.95 ($30.95 CDN)
ISBN 978-1-59327-137-4

The Unofficial LEGO® Builder's Guide
by ALLAN BEDFORD

The Unofficial LEGO Builder's Guide combines techniques, principles, and reference information for building with LEGO bricks that go far beyond LEGO's official product instructions. Readers discover how to build everything from sturdy walls to a basic sphere, as well as projects including a mini space shuttle and a train station. The book also delves into advanced concepts such as scale and design. Includes essential terminology and the Brickopedia, a comprehensive guide to the different types of LEGO pieces.

SEPTEMBER 2005, 344 PP., $24.95 ($33.95 CDN)
ISBN 978-1-59327-054-4

Steal This Computer Book 4.0
What They Won't Tell You About the Internet
by WALLACE WANG

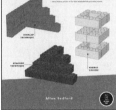

This offbeat, non-technical book examines what hackers do, how they do it, and how readers can protect themselves. Informative, irreverent, and entertaining, the completely revised fourth edition of *Steal This Computer Book* contains new chapters that discuss the hacker mentality, lock picking, exploiting P2P filesharing networks, and how people manipulate search engines and pop-up ads. Includes a CD with hundreds of megabytes of hacking and security-related programs that tie in to each chapter of the book.

MAY 2006, 384 PP., $29.95 ($38.95 CDN)
ISBN 978-1-59327-105-3

PHONE:
800.420.7240 OR
415.863.9900
MONDAY THROUGH FRIDAY,
9 AM TO 5 PM (PST)

FAX:
415.863.9950
24 HOURS A DAY,
7 DAYS A WEEK

EMAIL:
SALES@NOSTARCH.COM

WEB:
WWW.NOSTARCH.COM

MAIL:
NO STARCH PRESS
555 DE HARO ST, SUITE 250
SAN FRANCISCO, CA 94107
USA

companion website

Visit *http://www.thenxtzoo.com/* for updates, errata, and other fun stuff!